BIRD
Migration

BIRD
Migration

A New
Understanding

JOHN H. RAPPOLE

Johns Hopkins University Press

Baltimore

Johns Hopkins University Press
2715 North Charles Street
Baltimore, Maryland 21218–4363
www.press.jhu.edu

Library of Congress Cataloging-in-Publication Data

Names: Rappole, John H., author.
Title: Bird migration : a new understanding / John H. Rappole.
Description: Baltimore : Johns Hopkins University Press, 2022. | Includes
bibliographical references and index.
Identifiers: LCCN 2021002624 | ISBN 9781421442389 (hardback) | ISBN
9781421442396 (ebook)
Subjects: LCSH: Birds—Migration.
Classification: LCC QL698.9 .R375 2022 | DDC 598.156/8—dc23
LC record available at https://lccn.loc.gov/2021002624

A catalog record for this book is available from the British Library.

All illustrations not otherwise credited are the author's.

*Special discounts are available for bulk purchases of this book.
For more information, please contact Special Sales at specialsales@jh.edu.*

Among the most important discoveries for me concerning migratory birds is the central role played by the distaff members: not "subservient," not "subordinate," not "supporting," but *central*. Male dominance, long considered the wellspring of avian interactions between the sexes, is found to be impotent, at least during the migrant breeding season. It is females who call the shots, with all due respect to Konrad Lorenz and many others. Accordingly, I dedicate this effort to the extraordinary women who have played central roles in my own life: my wife, Bonnie Carlson Rappole; my mother, Francesca Goodell Rappole; my daughter, Brigetta Rappole Stewart; and my sisters, Francesca Rappole Wellman Miller and Rosemary Rappole.

Contents

Preface

Most people think they know a lot about bird migration. Almost everyone has seen many explanations for the phenomenon in a variety of media formats. However, these explanations are based on the premise that migration is all about weather (Figure P.1). It's not. Certainly weather, especially winds aloft, can be important in shaping some movements of some species some of the time, but it is not the ultimate cause.

Migration is about competition and seasonal change in food resources. It originates when young birds are forced to move away from the place where they were born in search of suitable habitat that is not already occupied by competitors—habitat where they can find mates and raise more young than they would have been able to had they stayed at home. This movement, known as "dispersal," is possible because of a superabundance of food. Without such a superabundance, the travelers would not get very far from home. Dispersal is the engine driving migration; seasonal change in food availability is the enabler. When food becomes superabundant— that is, available in greater supply than resident birds can consume due to seasonal climate change—the disperser can move and find new areas where it is possible to live and breed in the absence of, or at least in a reduction of, competition. If the climate changes at some later date, causing a disappearance of the foods on which the disperser depends, then the disperser must move again, and its best move is to go back to where it originated since that place, at least, is certain to have the proper habitat, albeit with plenty of competitors.

A critical caveat to keep in mind for the purposes of this discussion is the difference between climate and weather. "Weather" is defined as the physical characteristics of the environment—such as temperature, wind, and precipitation—at a given moment in time. "Climate" is these same physical characteristics over a period of time (months, years). Seasonal climate change, through its effect on food availability, is the driver of migration. Weather is not.

Attributing migration to dispersal from a more stable environment to a less stable one, with food and mates available on only a temporary seasonal basis, requires a complete reversal of how migratory bird ecology and evolution are currently understood. The breeding ground is not where the migrant originated from (it often originated from the wintering ground). Migrant evolution, including niche structure, is not shaped primarily by the breeding ground; instead, it is the wintering ground where migrants normally confront the most intense competition (having left a place where food was superabundant), and hence the most important aspects of natural selection occur here. Males do not control female behavior during the brief breeding season; in fact, the reverse is true due to the fact that females regulate male access to successful reproduction. The migrant annual cycle contains at least five seasons (spring migration, breeding, postbreeding, fall migration, and winter), not three (breeding, migration, winter). Each of these seasons has its own unique aspects and challenges that differ for the different age and sex groups and, indeed, for each individual based on its own specific experiences. And so on.

My purpose in writing this book is to explain migratory bird ecology and evolution from the perspective of an origin based on competition and dispersal rather than weather change. This approach requires a complete rethinking of the process—an exercise that is neither clear nor straightforward. In fact, the effort makes me mindful of a dream that I had recently in which I was sitting in a classroom with five or six fellow students and a professor. The prof said that before beginning the day's discussion, we would have a short quiz. He would state the questions, and we were to write the answers.

"First question: You are walking down the road. A scan ahead shows a stop sign. How far away is it?"

My first thought was "What!?" Then I thought, maybe a "scan" might be some archaic distance measure like a "league," but if so, of what length and in what system? If it is a measure, perhaps base ten, so maybe one hundred of something. Meters would be a lot for a pedestrian; maybe feet? So I wrote, "One hundred feet." By this time he was on to the next of about five questions, all similarly opaque to me. When finished, he asked for our answer sheets and said, "Let's get started." I raised my hand and said I was completely mystified by the first question. "What was the answer?" He looked to the young woman sitting next to me, who, with a slight smile, answered, "A tree." I just stared at her and then looked at the professor with my mouth open. He said, "Yes. That's correct. There's always a tree next to a stop sign." Then I woke up.

When I told my grandson, Zachary Stewart, who is a philosopher by nature and training, about this dream, he said, "You know, Poppy, a tree *is* a good answer, representing life at the terminus. But there's nothing wrong with your answer of a specific distance until the end is reached, either. In fact, probably the only wrong answer would be no answer." To which I responded, "That's all well and good, Zach. But what's it all doing in *my dream*?"

Thinking about this, I realized that my situation in the dream was similar to that in which many people find themselves when given an explanation for a particular phenomenon; they are made to feel that only initiates can even understand the questions properly, let alone the answers. I think that I was guilty of this approach to some degree in my book *The Avian Migrant*, published in 2013 by Columbia University Press, wherein I attempted to summarize current understanding of the principal concepts of bird migration based on dispersal as origin. I had hoped that the book would be read and understood widely by interested lay people as well as experts, just as a lot of folks who are not astrophysicists read Stephen Hawking, but that has not been the case. In addition to the "dream" problem of more or less willful obfuscation, the requirements of scientific writing are at odds with what constitutes a readable narrative for the person who comes to the topic solely out of interest. In a scientific publication, any statement of fact must either be documented by intrinsic data or references containing such data. Specialists reading a scientific paper expect this type of documentation, and lack of it is considered weak reasoning. Nonspecialist readers, however, find in-text citations to be distracting to say the least.

Another issue is that, in a technical manuscript presenting new findings, the author must assume that readers bring a vast knowledge of the field with them. Otherwise, sections addressing each topic would be book-length in themselves.

I know that these factors were problems for general-interest readers of *The Avian Migrant* because several of my friends told me they found it mostly unintelligible. In fact, my Columbia University Press editor had to search hard for professional reviewers for each separate chapter because one who might consider himself or herself an expert in one aspect of migration, say physiology, would feel unqualified to critique another, say population ecology. If that is the case for specialists, then it is no wonder that my friends had trouble, even though they had a great interest in the topic.

One should not have to know arcane languages and symbols in order to understand the basic concepts for any major piece of objective truth. It is up to the person presenting the arguments to make them comprehensible. It is certainly true that migration, like most aspects of our life on earth, has elements that are unknown; indeed, the "mystery of migration" is a well-worn but still apt cliché. Nevertheless, the topic deserves a clear account that is understandable by any interested reader.

My hope is that people who just want to know about migration will pick up this book and find it difficult to put down. I want to engage them in what has been for me the pivotal and absorbing pursuit of a lifetime in science. Nevertheless, many of the concepts addressed are complex, requiring significant focus if the reader is to obtain any meaningful level of comprehension.

Narratives do not lend themselves easily to the precise, empirical arguments typical of scientific works. In fact, in taking this approach, I am indulging myself in a fair amount of speculation, which, of course, has a bad reputation in science because too often it is presented as fact when it's just a WAG (wild-ass guess). Whole disciplines have been built on WAGs, although the guess part is often hidden semantically behind multisyllabic terms like "heuristics" and "hermeneutics." I freely acknowledge that, in presenting the following summary of the field of migration as a chronicle, I am engaging in wholesale mixing of fact and guesswork, because there is still a great deal that we do not know.

The story of bird migration involves some of the deepest mysteries of our existence, and in telling it, I am going to have to resort to some involved arguments. Also, I am going to have to use some jargon. Without it, technical discussions quickly become unwieldly. Nevertheless, I promise to define my terms, like a good logician.

It is obvious to most people that migration originated as a result of northern winters; bad weather simply forced birds to head south until spring. Thus, the central thesis I present herein, that many migrants originated as tropical natives, is absurdly contrary to common sense (and existing paradigms). In that, however, the theory is in good company. Many, perhaps most, fundamental truths of science began their existence as counterintuitive fables, including the fact that the earth revolves around the sun.

I begin this book by discussing the basic tenets of the most widely accepted explanation at present for how migration works, the Northern Home Theory. I then present information gathered during field research on the various parts of the migrant annual cycle by me and many others, along with a reconsideration of our understanding of migration based on a new hypothesis, the Dispersal Theory, which is forced by these data. Finally, I consider migrant population dynamics, evolution, biogeography, and conservation as viewed from a Dispersal Theory perspective.

Birds are the main actors in this drama, and, like the man says, "You can't tell the players without a program." Herein, I will use only the common name for each species in the text, following the American Ornithological Society's *Checklist of North and Middle American Birds* (http://checklist.aou.org/taxa) as the reference guide to nomenclature for all birds native to those regions. Sources are trickier for the rest of the world. I will use *Birds of the World* by Frank Gill and Minturn Wright for the most part, except in cases where they are not in agreement with the names used in the particular work I happen to be discussing. In those situations, I will use the names used by the author in question. All species mentioned in the text are listed by common and scientific name in Appendix 1.

Migration is an incredible journey. Why does it happen? How is it possible? My hope is that by the time you have completed this book you will have gained some idea of what we know about the answers to these questions, as well as what remains to be discovered, and that you will have enjoyed the trip.

Acknowledgments

Now praise we great and famous men [and women!].
—Hymn by William George Tarrant

Honor is a major theme of *The Iliad*. Therein, Homer takes great care to present the names, homelands, and feats of his heroes, which is why Robert Fitzgerald in his wondrous translation of the poem renders those names in as close an approximation as he can to the ancient Greek instead of some foreign bastardization (who the heck is Aias anyway?).

Honor is important in most facets of life, and this is certainly true for scientific investigation. Indeed, seeking such honor is an important reason for following the calling for many of us, which is one reason why significant care should be taken by authors in presentation of data and ideas of others. We owe it to those who laid the foundation.

Too often such debts are ignored. A dear friend and long-time collaborator, Czech virologist Zdenek Hubálek, recently contacted me with a complaint: "I read a new paper on West Nile virus that essentially presented our ideas on how the disease moves as their own. They didn't even bother to cite us. Is this the way they do things in your country?"

Sadly, I had to admit that it is the way "they do things" in my country, and a lot of others, unfortunately. If reviewers and editors do not hold authors to the basic ethics of science in properly crediting the sources of their ideas, then many authors will be quick to publicly appropriate the honor of discovery for what is not theirs. If I have done that in my published work, it is through ignorance. A narrative provides little space for detailed recognition, but I attempted to be meticulous in this regard in my book *The Avian Migrant*. Readers interested in the details of who found out what and when concerning the principal building blocks of migration science should consult that source.

Of course, some debts are too large to be passed off in such a way. The purpose of this section is to recognize those who made major contributions to my thinking and career and, in particular, those on whose thought and effort the Dispersal Theory of migration origin was built. Following the Homeric tradition, I list the principal contributors below, along with their deeds.

Heading the list, of course, is Charles Darwin. The entire field of evolutionary biology grew largely from his thought, but there are two aspects in particular that apply specifically to development of migration theory. The first is his recognition of the importance of instinct as a biological trait subject to natural selection. Others before him, notably Jean-Baptiste Lamarck, had observed that animal structural traits appeared to be crafted in some way by the environment, but Darwin made clear in his work that behavioral traits were similarly subject to the laws of natural selection. He used artificial selection as an example of this process at work. Dogs, for instance, came to us as wolves with the full panoply of wolf-like behaviors. Humans treated these behaviors as a palette and quickly modified wolf-like behavior through selective breeding into a bewildering variety of types, each with its own particular behavioral specialty: pointers, herders, defenders, fighters, lurchers, racers, trackers, fetchers, and so forth.

The second major contribution of Darwin toward our understanding of migration was his singling out of sexual selection as a special form of natural selection in his 1871 book, *The Descent of Man, and Selection in Relation to Sex*, wherein he made clear that the selective environment was quite different for males and females of the same species with commensurate results in terms of their different structures and behaviors.

Second on my list of heroes is Arthur Landsborough Thomson. In his book *Bird Migration*, published in 1936, Dr. Thomson succinctly summarizes inscrutable facts regarding migration, several of which continue to baffle us today. Margaret Morse Nice, David Lack, and Val Nolan Jr. performed similar services for those trying to make sense of the migration phenomenon with their wonderful insights derived from seminal work on the Song Sparrow, European Robin, and Prairie Warbler, respectively. Rice University English professor George G. Williams, despite his lack of ornithological credentials, had the temerity to challenge the establishment on the issue of avian migration route flexibility decades before eBird (the online source for North American bird distribution). Jerram Brown, Steve Fretwell, and Alan Pine provided the theoretical underpinnings for an understanding of how competition, the starting point for any understanding of migration, works during the various stages of the migrant life cycle. Eberhard Gwinner, along with his colleagues and students at the Radolfzell Lab in Germany, demonstrated that migration timing was programmed in different species of birds in different ways. His student Barbara Helm, through her rigorous testing of ideas developed in the lab, greatly expanded our understanding of migration, while helping direct focus to mysteries yet defying explanation. Robert MacArthur established the ecological and behavioral parameters to look for in migrant populations if the Northern Home Theory were correct. Arie van Noordwijk developed the idea of "reaction norms," a powerful concept that explains the fluidity of presumably fixed and genetically controlled migrant behavior under differing environmental conditions. Theunis Piersma and colleagues took the "exaptations" concept, proposed by Stephen J. Gould and Elisabeth Vrba, and used it to explain how whole suites of characters assumed to be specific to migrants could, in fact, have originated for other purposes entirely, thereby granting a new conceptual freedom with regard to understanding how migration comes about.

The people discussed above, along with many others, paved the way for those of us who continue to investigate migration. The following account credits those with whom I have worked directly in my research career on migrants, which has spanned the past forty years. The following colleagues shared as principal investigators on research projects: Patricia Escalante, Mara Neri, Sherry (Pilar) Thorn, Brad Compton, Peter Leimgruber, J. Robertson,

Tim Fulbright, Jim Norwine, Ralph Bingham, Dave King, Swen Renner, Dick DeGraaf, Scott Derrickson, Zdenek Hubálek, Selma Glasscock, Ken Goldberg, Dezhen Song, S. Faridani, Barbara Helm, Mario Ramos, Peter Jones, Arlo Kane, Rafa Flores, Alan Tipton, Wylie Barrow, Jeff Diez, Jorge Vega Rivera, Vicki McDonald, Gene Morton, Tom Lovejoy, Jim Ruos, Alan Pine, Dave Swanson, Gary Waggerman, Dick Oehlenschlager, Dwain Warner, J. F. Hernandez Escobar, Chris Barkan, Kevin Winker, Charly Schuchmann, Bill McShea, George Powell, John Klicka, and Steve Sader.

The majority of my research projects involved a great deal of help in the field, mostly provided by family, friends, and students, but often enough including recruits from among our neighbors in whatever place we happened to be working. It is my pleasure to honor these long-suffering assistants: Bonnie Rappole, Chris Barkan, Bruce Fall, Bob Zink, Mario Ramos, Carol Gobar, Sherry Thorn, T. Holt, O. Iglesias, J. Leiba, M. Martinez, D. Menendez, Joan Milam, J. Miguel Ponciano, J. F. Hernandez, many volunteers from the Earthwatch program, Rebecca Scholl, Peter Leimgruber, K. Korth, Liz Reese, T. Small, D. Fletcher, P. Cross, Angel Toto, Refugio Cedillo, Judy Wilson, John Barber, Joe Folse, Becky Bolen, Karilyn Mock, Howard Galloway, Doug Mock, Isabel Carmona, Jose Luis Alcantara, Mara Neri, Samantha Baab, Isabel Castillo, Rick Coleman, Dave Delahanty, Kevin Winker, Doug Gomez, A. Hartlage, Jacinto Hernandez, John Howe, John Klicka, Ben Robles, Steve Stucker, Erma (Jonnie) Fisk, Courtney Conway, Patricio Choc, Walter Van Sickel, Harriet Powell, Brian Miller, Isabel Valdovinos, Lisa Yoder, Mike Yoder, Martha Van der Wort, Marcia Wilson, Andy Wilson, and Nellie Tsipoura.

Field work is expensive and requires a lot of help, in terms of permits and information from various government and private organizations, in order to be successful. In the following list, I credit organizations for their investment in my work: World Wildlife Fund (WWF), Instituto Nacional de Investigaciones sobre Recursos Bióticos (INIREB), Hewlett-Packard, Esri, ERDAS Inc., Honduran Forestry Department (COHDEFOR), the Guatemalan National Protected Areas Council (CONAP), the Mexican Secretariat of Environment and Natural Resources (SEMARNAT), US

Department of Defense (Legacy Program), Smithsonian Institution (Scholarly Studies program, Atherton Seidell Endowment Fund), Smithsonian National Museum of Natural History (Alexander Wetmore Fund), Smithsonian National Zoological Park, Friends of the National Zoo, American Museum of Natural History (Frank M. Chapman Memorial Fund), Welder Wildlife Foundation, National Fish and Wildlife Foundation, the Center for Field Research, the US Forest Service, Shenandoah National Park, USGS Bird Banding Laboratory, Departamento de Fauna Silvestre (Mexico), Universidad Nacional Autónoma de México (Instituto de Biología, Minnesota Department of Natural Resources, Georgia Department of Natural Resources (Nongame Division), Texas Parks and Wildlife Department, US Fish and Wildlife Service (Migratory Bird Office), Caesar Kleberg Wildlife Research Institute, US Department of Agriculture (Animal Damage Control), Smithsonian Migratory Bird Program, WildWings Foundation, USAID, Inter-American Development Bank, National Biological Service, Defenders of Wildlife, German Academic Exchange Service, and the Garfield Foundation.

Figuring out whether a manuscript has any value, and, if so, what is wheat and what is chaff, is not an easy or rewarding job. I thank those who undertook the Augean task for me in reviewing this work: Richard Chandler, Jim Berry, Mary Ellen Vega, Gene Blacklock, and Kevin Winker.

Finally, I thank my long-suffering editor at Johns Hopkins University Press, Tiffany Gasbarrini. The fact that she stood with me through the seemingly innumerable revisions and complete shifts in direction is a testimony to her vision and dedication. This work looks very different from the initial version I sent in back in early 2018, and I credit her for any coherence it now appears to possess.

Mistakes, errors, goofs, failures, gaffes, and so forth are entirely the responsibility of the author.

CHAPTER 1

THE BIRD
Migration Paradigm

Suspect your faith in the writings of the ancients and submit it to experiment.
—Ibn Alhazen, *Book of Optics*

The guiding principle in this book is that migration is a form of movement—originating as dispersal and enabled by a superabundance of food—by individuals away from their place of birth or population of origin to a seasonally suitable breeding area and a return from that area to their point of origin on the disappearance of food.

The number of bird species that are migratory depends on your definition of the term "species." Ernst Mayr, author of the biological species concept, stated that a species is "groups of actually or potentially interbreeding natural populations reproductively isolated from all other such groups." Following that definition, which has been accepted for the past half century, there are in the neighborhood of nine to ten thousand bird species in the world. Consensus has shifted somewhat, however, and, using new insights regarding what constitutes a species, there may be as many as eighteen thousand bird species.

Similarly, our understanding of migrants and migration has changed, making an estimate of numbers problematic. Back in 1995, for my book *The Ecology of Migrant Birds*, I calculated that there were 338 species (using

Mayr's species definition) that breed in north temperate or boreal regions of North America and winter all or in part in the New World tropics; 185 that breed in Europe and winter in Africa; and 336 for temperate and boreal Asia that winter in the Asian or island tropics. The numbers are much higher if one includes short-distance migrants; indeed, most birds that breed north of the tropics migrate to some degree. The same is true for many species that breed south of the Tropic of Capricorn, although the extent of these movements is only now becoming clear as countries from these regions institute banding programs. Even less well-understood are intratropical movements. The majority of the world's bird species are indigenous to the tropics (greater than seventy percent). Migration has been documented in only a few of these, mostly fruit- or nectar-eaters that undergo seasonal movements tracking their food resources, but the actual number likely is much higher.

When I began my field work as a graduate student at the University of Minnesota's Bell Museum of Natural History in the early 1970s, the scientific consensus concerning bird migration was that avian migrants breeding north of the Tropic of Cancer in the Northern Hemisphere ("Nearctic" for short) originated as temperate and boreal zone residents that were pushed southward each year at the close of the breeding season by deteriorating weather, gradually evolving over millennia the physical, physiological, and behavioral adaptations for a migratory way of life that we see today.

By the '80s, the data that we had collected on migrants during the breeding season (in Minnesota, in transit in Texas, and on their wintering areas in southern Veracruz), as well as field work done by others, convinced me that this Northern Home Theory did not adequately explain vast portions of the Nearctic migrant life cycle. A new hypothesis of wintering-ground origin that better fit the data for most migrants needed to be put forward and tested, namely that migrants were not interlopers where they wintered but had their own ecological space (niche) in these environments; that is, they were tropical birds that headed north to breed rather than temperate birds that headed south to winter.

At present, a new iteration of the Northern Home Theory, known as the Migratory Syndrome, is the most widely accepted idea among biologists for the origin and evolution of migration. It is a concept developed by the Australian entomologist V. Alistair Drake and colleagues in a book published in 1995. Therein, the syndrome is described as "a suite of traits enabling migratory activity." The concept was developed for application to insect migration, but it has been widely applied to other faunal groups. As it relates to understanding the origin of bird migration, the "syndrome" can be described as an ancient gene or attribute (several genes) that controls a highly integrated bundle of adaptive traits—including such things as hyperphagia (intensive eating), migratory flight, navigation, orientation, etc.—generally assumed to have evolved over eons of selective pressure, likely imposed by gradual change in the climate from warm and more or less aseasonal to cooler and seasonal. In some versions, the gene or bundle is considered to be present in all or nearly all birds due to its possible origin in dinosaur progenitors, or an even earlier origin. This variant of the theory often includes the idea of a genetic "switch" that can be turned on and off by environmental conditions and pressures. In other forms, the bundle has evolved in several avian lineages at different times and places, but also many millions of years in the past. The idea is that this suite of characters (genetically inherited physical or behavioral attributes), once evolved, is highly conservative, perhaps like web-making in spiders. It will always be there in derivative phylogenetic linkages, although perhaps suppressed or expressed in different ways under different environmental conditions.

Ironically, if one substitutes the term "dispersal" for "migratory" as the descriptor for the proposed syndrome, it becomes very much like what is presented in this book as the basis for the origin of migration. Dispersal is an evolved strategy likely common to every creature on earth, and, although certainly it will differ from group to group and species to species depending on the capabilities and life history of the organism, dispersal will indeed involve a suite of characters whose evolution dates back in deep time to the earliest forms of life. No "switch" is needed, however, to trigger dispersal. It's a package always ready for use when needed, as it almost always is for young individuals of density-dependent organisms shortly after reaching independence from their parents.

A recent expression of the Migratory Syndrome hypothesis for birds was presented by Benjamin Winger and his associates in the 2014 *Proceedings of the National Academy of Sciences* wherein they credited me as an author of the "dominant paradigm" for the origin of bird migration via dispersal. This characterization is ironic, to say the least. You could probably fit all of those who acknowledge the validity of the Dispersal Theory for migration origins into a small dining hall with plenty of room left over for the caterers. Be that as it may, I accept the designation, at least when enclosed in quotation marks. Although there is much that is not new about the Dispersal Theory, it is certainly different from both the Northern Home and Migratory Syndrome hypotheses.

The *Oxford English Dictionary* defines "paradigm" to mean "a typical example or pattern of something." It is only recently that the term has assumed its current pejorative sense, mainly due to the writings of the historian of science Thomas Kuhn, who defined a paradigm as "a set of concepts that govern theoretical understanding for a given scientific area of inquiry at any particular period of time." This definition seems harmless enough, except that it contains within it the implication that the "set of concepts" will be shown to be wrong by the accumulation of inconvenient data at some later period. As Kuhn explains, such a set often dominates the thinking of practitioners in any given field who come to believe it to be reality rather than a working hypothesis, and who treat those testing it as unworthy of hiring, publication, and funding (or even as heretics, as in the case of Galileo and many others).

The Migratory Syndrome, in its most recent "molecular systematics" form, is built from the top down, which is to say that understanding of migration is derived from what molecular geneticists call "deep history," dating back to the dawn of flight, or even before. The Dispersal Theory, as presented herein, is built from the bottom up—the idea being that the way that organisms are structured and behave tells you how and why they do things, including migration.

This notion has been discredited in much recent discussion, the perceived problem being that when you look at a behavior or a structure, you only see a smidge, indeed, an insignificant part of the puzzle. It is true that those of us who spend our days making observations often abuse Occam's razor—that is, we tend to build entire theories from such scintillas. Nevertheless, critics of field work on this basis often forget that while a single observation is no more than an anecdote, a number of observations is a data set.

The phylogenetics version of the Migratory Syndrome makes no mention of how a high-latitude origin for migration might affect our understanding of the ecology and population dynamics of migrants when compared with a lower-latitude origin. In this aspect, it is quite different from other variants of the Northern Home Theory, which, in fact, built vast theoretical underpinnings regarding migrant evolution and ecology, beginning with the assumption that migration originates when birds are pushed south by weather, and then, in Procrustean fashion, retrofitting understanding of the migrant's life history into this mold.

When I began my migration studies, ecological thinking and theorizing were dominated by Robert MacArthur, a fact which was as true of bird migration theory as of many other areas of ecology. MacArthur's migration paradigm states that migrants are temperate zone birds forced south (north for Southern Hemisphere species) by inclement weather into warmer climes for the duration of the winter season, where they subsist in marginal environments by harvesting superabundant food resources that resident bird species cannot exploit entirely due to the restricted nature of these foods in space and/or time. Migrants then return to the breeding grounds once the weather has ameliorated. This concept was based largely on ad hoc observation of songbird migrants wintering in Mexico and Central America, where such birds often can be seen in scrubby second growth of fallow croplands and pastures as well as in parks and residential areas of towns and cities.

Perhaps the best description of this phenomenon as it actually happens in the natural world came from work on wintering European migrants in the sub-Saharan region of Africa, particularly as presented in a paper by French biologists Gerard Morel and François Bourlière in 1962, where they described these birds as "floating populations" of homeless wanderers subsisting on "superabundant resources" that could not be harvested by resident species because of their temporary (hours? days? weeks?) nature.

Strictly speaking, this Northern Home idea of bird migration is not MacArthur's paradigm. He did not originate this theory. In fact, the concept goes as far back as published thinking on the topic exists, more than two hundred years. However, MacArthur's contributions to it are critical. In typical fashion, he made logical, testable predictions regarding aspects of migrant biology that followed naturally from the central assumption of a breeding-ground origin. Among these predictions were the following:

- The true niche for a migrant—that is, the ecological space including food and habitat unique for its species—exists only on its breeding grounds.
- Once the migrant leaves its breeding grounds, it is outside its niche, which means that it must survive as a wanderer (fugitive species).
- Stable environments in areas outside the breeding grounds will be occupied by indigenous species, and in these environments, species similar to the migrants will exist as "ecological counterparts," where they will fill niches comparable to those filled by the migrants on their breeding grounds.
- Ecological counterparts prevent migrants from settling for long periods (weeks or months) in stable environments either directly, by chasing and fighting them, or indirectly, by outcompeting them for food.
- Therefore, migrants exploit mainly marginal environments, such as overgrown fields or brushy areas, in their wanderings during the non-breeding period, searching for patches of temporarily superabundant food that resident species are unable to harvest completely.
- Entry by migrants into more stable environments in the subtropics or tropics will occur only in situations where resources (such as fruiting trees or army ant swarms) are temporarily so abundant that they cannot be harvested completely by, or defended by, resident ecological counterparts.

MacArthur's predictions make clear the dilemma confronting a temperate zone migrant. The entire suite of structural, physiological, and behavioral adaptations for this bird supposedly evolved millions of years ago, according to Northern Home theorists, when its high latitude homeland was more climatically benign throughout the year. Now, these adaptations are no longer suitable for survival during northern winters. Thus, each fall, the migrant "niche" (ecological space unique to each species) disappears for months at a time, and the bird is forced into more southerly regions where all of the niches are already filled by a resident community of birds.

Resolution of this paradox is what the remainder of the book is about.

CHAPTER 2

THE MIGRANT
Annual Cycle According to
the Dispersal Theory

What songs the Sirens sang, or what name Achilles assumed when
he hid himself among women, although puzzling questions are not
beyond all conjecture.

—Sir Thomas Browne, *Hydriotaphia*

In 1979, I was awarded a contract to prepare a book summarizing
literature on Nearctic avian migrants in the neotropics. This was a brain child
of Gene Morton, research scientist at the Smithsonian Institution, and Tom
Lovejoy, vice president for science at the World Wildlife Fund, with funding
from a partnership between the World Wildlife Fund and the nongame section of the US Fish and Wildlife Service, headed by Jim Ruos at that time.

This contract allowed me, with the help of a part-time editorial assistant (Candy Yelton in 1980 and Linda Kundell in 1981), to conduct an
exhaustive review of the existing literature on migratory birds. Under the
auspices of this assignment, I spent two years attempting to find and read
every published account on the topic of bird migration in general and New
World migrants in particular. Since this was before the days of the internet, the main methods involved a computer literature search through the
resources of the excellent Science Library of the University of Georgia,

followed up by extensive copying of articles from the library's journal collection, interlibrary loans, and correspondence. Eventually we compiled a cross-referenced and annotated bibliography of over 3,300 citations.

The information gleaned from these publications, in combination with my own research findings in Texas, Mexico, Minnesota, and Georgia, afforded exceptional insights into the phenomenon of migration, which can be summarized as follows:

- More than 330 of the 650 species (fifty-one percent) of birds which make up the avifauna of the continental United States are Nearctic migrants whose populations winter all or in part in the neotropics.
- Populations of these species spend one-half to two-thirds of their life cycle outside of the United States in tropical communities.
- Every major habitat type in the neotropics supports some Nearctic migrants; 120 species inhabit shrub-steppe environments, 105 species winter in aquatic habitats, while 107 live primarily in various types of neotropical forest.
- The percentage of migrant species in tropical forest communities varies from a low of four percent in montane wet forest of Colombia to fifty-one percent in the gallery forest of Colima along the western coast of Mexico.
- Migrants, like many resident species, can also be found in various types of second growth. Four explanations are suggested for this behavior:
  a. Resemblance of certain seral (intermediate) stages of development in some habitats to mature stages of other habitats.
  b. Resemblance of some facet of seral stage structure to various aspects of mature habitats.
  c. The presence of temporarily abundant, easily harvested food.
  d. The second growth is undefended, unlike the preferred habitat.
- Migrants have broad dietary preferences and many species change their diets according to what is available during different seasons.
- Of the ten major food-use categories, the largest percentage of migrant species (eighty-two percent) include arboreal invertebrates in their

diet, followed by aquatic invertebrates (thirty-eight percent) and fruit (thirty-two percent).

- Migrants are not excluded by residents from mature tropical communities or the stable resources of those communities; many show long-term fidelity, remaining on a site throughout a winter season and often returning to the same site in subsequent seasons.

- Many species of migrants defend winter territories against conspecifics (members of the same species) as a means of assuring limited resources for their own use. This sort of intraspecific competition (between members of the same species) indicates that food can be scarce during the winter season, and that many migrant populations are limited during such times.

- Nearctic migrants that winter in the neotropics, like residents, function as integral members of the communities they inhabit.

- Migrants of some species attend multispecies flocks in tropical forest. The main benefit derived from this behavior is most likely related to predator avoidance (more eyes keeping watch).

- As members of neotropical communities, migrants affect community structure and function in subtle ways. One of the best-documented of these effects relates to fruiting periods of tropical trees, some of which peak during the spring passage of transients.

- Forty-eight percent of all Nearctic migrant species have populations of conspecifics that breed in the neotropics. Seventy-eight percent have either congeners (members of the same genus) or conspecifics that breed in the tropics. These figures, along with the data on wintering ecology, indicate that many Nearctic migrant species are essentially tropical birds that evolved the habit of migration as a means of periodically exploiting temperate habitats in which food was more abundant, competitors fewer, predation lower, and reproductive success (number of offspring produced) greater.

- Nearctic migrants depend on tropical communities for survival. The near extinction of Kirtland's Warbler and Bachman's Warbler and the decimation of the Golden-cheeked Warbler may be examples of what effect wintering-ground habitat destruction can have on migrant populations.

- There is evidence from several sources to indicate that US breeding populations of certain Nearctic migrants are declining. The most probable reason for declines in most migrants is destruction of wintering habitat.
- All tropical habitats are threatened by rapid human population growth and uncontrolled development.
- Our best hope for preserving our Nearctic migrant birds is to accept the biological reality that they constitute an internationally shared resource that can be protected only through cooperation in policy formulation, research, and management.

These understandings provided further evidence for me that the Northern Home Theory of migrant origin simply did not fit the known facts of migrant ecology and taxonomic relationships.

Perhaps the best expression for the Northern Home Theory so far as it applies to migrant wintering ecology in the tropics is that proposed by Edwin Willis. He dubbed this explanation the "irregularity principle," which he defined as follows: "Biological or physical irregularities in the environment create open niches or superabundances of food which are available to wandering populations of wintering migrants." This hypothesis, despite its name having an unfortunate resemblance to a malady best treated with a laxative, was a brilliant reformulation of MacArthur's fugitive species hypothesis to fit known facts—namely, that migrants were not excluded from mature, stable tropical habitats like rain forest. Willis was a superb tropical field biologist (he died in 2015) whose main interest was in the biology of ant birds, a group of tropical species specializing in a symbiotic relationship with army ant swarms wherein, by tracking the swarm, the birds are able to capitalize on capture of prey dislodged by the swarm's movement along the forest floor. Willis spent thousands of hours observing these interactions in forests of many parts of the neotropics, and he knew that some species of migrants often were regular attendants at swarms in primary forest, contrary to MacArthur's prediction that they would be excluded by tropical resident ecological counterparts. His irregularity principle expanded the idea of temporary food superabundance to include the forest interior where ant swarms, tree-falls, seasonal fruit production, and so forth might explain how migrants could exploit resources in these mature habitats despite the presence of resident ecological counterparts.

Validity of the irregularity principle is essential to a Northern Home Theory of migrant origins because it explains how members of species that are pushed southward by weather can survive the winter without the need of a niche (ecological space) that is uniquely their own in tropical environments. If food resources are superabundant, then there is no need for a niche.

It may help to consider this critically important concept through use of an analogy. Think of birds at a feeding station abundantly supplied with suet and seed. So long as there is plenty of food, it doesn't matter who chases whom; all members of all species who know how to use the station will get enough to eat. What Willis is saying is that an ant swarm or a fruiting forest tree is like a feeding station where there are more resources available than members of resident species can exploit, so migrants can live in mature tropical forest environments as long as the excess lasts. The key aspect of this theory of migrant resource use during the nonbreeding period is "irregularity." According to Willis, and MacArthur as well, these resource pools are irregular in space and time, which means that migrants have to be continually on the move throughout the migration and winter seasons to find and exploit these temporary superabundances, hence the idea that migrants must be wanderers during these times.

The importance of our field work in the Tuxtla Mountains of southern Veracruz was that we found that many individuals of several species of Nearctic migrants in forest did *not* move. Through banding, recapture, and resighting, we found that individuals of at least twenty species of migrants lived on small pieces of ground in mature rain forest habitat throughout the winter season, and that many returned to these same plots the following season at rates as high as sixty percent. Clearly this discovery is contrary to the displacement from stable habitats predicted by the irregularity principle.

Nevertheless, migrant winter residence in primary tropical habitats could be explained by a situation in which resources remain superabundant for all of the migrant species throughout their winter residence.

It is interesting to note that situations in which long-term persistence (weeks or months) of periods when more food resources are produced than can be consumed by indigenous species are, in fact, critical to our understanding of why migration is such a successful strategy for hundreds of bird species—because resource superabundance is a characteristic of many higher latitude regions of the world during late spring and summer months. Resource superabundance is also characteristic of the sub-Saharan region of Africa (and a number other places in the world) for a month or two in November and December, the time period during which Morel and Bourlière made their observations of floating populations of in-transit migrants in Senegal.

At our study sites in the Tuxtlas, this condition of persistent resource superabundance in primary habitats was contradicted by two observations: territoriality and wanderers. By definition, territoriality occurs when critical resources are limited, consequently requiring defense. Wanderers (also known as "floaters") are nonterritorial birds that move through the plots of territory owners. They will rapidly assume ownership of the territory if given the chance, as described for Hooded Warblers on our Veracruz study sites. The presence of such individuals further demonstrates the limited nature and desirability of rain forest territories for some migrant species. The existence of several species of migrants defending territories throughout the winter season in stable, tropical environments like rain forest means that the Northern Home Theory, in which migrants originate as temperate or boreal residents pushed south over millennia each fall as interlopers into south temperate or tropical environments, is invalid, at least for these species.

But what is the alternative? Obviously, the Dispersal Theory, in which migrants originate as tropical residents that move north in summer to exploit seasonal food resource superabundances in northern regions. The finding discovered through my literature search, mentioned above, that forty-eight percent of long-distance migrants to the tropics have populations that are resident in the tropics, lends powerful support to this hypothesis.

The alternative, following the Northern Home Theory, would be that these interlopers simply pushed their way into tropical habitats during the winter period and stayed on to breed during the summer, becoming tropical year-round residents. This idea, referred to as "migratory drop-off" by its proponents, is actually at the heart of the hypothesis put forward by Winger and company, reviewed in detail in Appendix 2 and discussed in the chapter on origin and evolution (Chapter 9).

Most people are surprised when confronted with the fact that nearly half of long-distance migrants have breeding populations in the tropics, and some are openly disbelieving. But then, their idea of what constitutes a long-distance migrant is often limited to a few kinds of birds, say Baltimore Orioles, Scarlet Tanagers, and Red-eyed Vireos. They don't think about the hundreds of species that make up this class, including waders, raptors, waterfowl, hummingbirds, cuckoos, doves, swallows, swifts, gulls, sandpipers, nighthawks, and kingfishers, in addition to most songbirds. This perception is also true of many ornithologists, who, when they think of migrants, tend to think of thrushes, flycatchers, warblers, vireos, and the like that breed in temperate and boreal forests. Cursory surveys of tropical habitats show individuals of many of these species to be common in marginal and otherwise disturbed habitats, such as residential areas, pasture borders, and scrub, which is what makes the Northern Home Theory so appealing. It is obviously true. These birds are clearly forced by winter weather to go south, where they wander about subsisting on patches of superabundant resources in marginal tropical and subtropical environments, prevented by residents from using the stable resources, such as those in rain forest. Except that they aren't. It is true that they do head south, and many of them do end up wintering in marginal habitats, but that does not necessarily mean that they are pushed south by weather, particularly since most migrate long before the weather becomes problematic. Also, it does not mean that they are prevented from using stable primary habitats by residents. In fact, they are prevented by territorial members of their own species in many cases. In any event, understanding how these birds fit into a Dispersal Theory was a major part of my research focus for several years.

By this time, you are probably wondering what the heck I was thinking of in providing quote at the beginning of the chapter. I love Sir Thomas Browne's use of the word "conjecture," which seems to fit quite well with my intent in the following chapters on the key parts of the migrant annual cycle reimagined from a Dispersal Theory perspective. What Sir Thomas is saying, in the passage from which this quote is taken, is that Sirens' musical tastes and Achilles' experiences hiding himself are things about which we *can* speculate, whereas the bones discovered in ancient Grecian burial urns can tell us nothing whatsoever about the rich lives of the people to whom they originally belonged—these people are gone forever and no speculation regarding them is possible.

This dictum seems to me to be a good takeoff point for introducing the remainder of this book. Therein I provide the reader with the information gathered from personal research and the literature on various aspects of migrant ecology. These data raise questions about which it is perfectly valid to speculate—but the reader should keep in mind that speculating is what we are doing. These conjectures are no more than working hypotheses, and that's what much of the rest of this book is about—call it exegetics, hermeneutics, heuristics, or WAGs (wild-ass guesses)—it all amounts to the same thing. I have some ideas concerning migration which I am going to present to you, but you should not take these as truth.

What few people understand about scientific narratives is that they share the same fundamental conceit as an historical novel; that is, the novelist employs known facts as his takeoff point for his flights of fancy. Science writers do the same, although they often attempt to blur the distinction by mixing data with WAGs. From this point forward, I will follow this path. Consider yourself warned. If you want a more careful separation of fact from fancy, you can always refer to *The Avian Migrant*.

It took me a very long time to understand why there was so much resistance in the scientific community to the idea that the Northern Home Theory of migration origins simply did not fit the facts. It only really sank in for me when I was writing the annual cycle chapters for this book. It was then that I finally realized that nearly the entire field of ornithology is based on the assumption that the breeding period is the key to understanding virtually all aspects of avian ecology and evolution. This approach is made very clear by looking at the vast majority of summaries of avian life histories, from those written in the early 1800s (like Alexander Wilson's three-volume tome, *American Ornithology*) right up to the American Ornithologists' Union's *The Birds of North America*, edited by Alan Poole and published in eighteen volumes between 1992 and 2015, where the vast majority of information presented in the accounts for the seven hundred plus species treated concerns the breeding period. Migration and wintering information is practically nonexistent.

The inescapable conclusion from these types of presentations is that the breeding ground is the only place where anything of importance happens to the bird. Therefore, all structures, behaviors, and physiological attributes must derive from this period of the life cycle. The Northern Home Theory fits this interpretation of migrant life cycle function perfectly. According to this theory (and the Migratory Syndrome), the birds evolved on the breeding ground. When they are pushed out in the fall, they are moving to places where there is more food available than the residents in those places can harvest. Therefore, the invading migrants don't need any special adaptations to compete. They are just like migrants visiting a bird feeder in winter. Sure, they might get pushed around a bit by the birds that are there all year, but so what? There is plenty of food for all. Thus, with no death occurring as a result of competition, there is no evolutionary change occurring that is worth mentioning. Migrants are simply wandering from one resource cache to the next until it is time to head back north where the real action is. However, if the Dispersal Theory is correct, this turns our entire understanding of the migrant life cycle on its head. It is the wintering ground where competition is most intense and the breeding ground where food is superabundant.

Now you may see why I have quoted Sir Thomas. The following sections on the migrant life cycle contain significant "conjecture" as I strive to put what we actually know about these periods into the context of a dispersal origin for migration.

CHAPTER 3

FALL
Migration

Fall migration is caused by weather. The leitmotif of this book is that this hypothesis is false—that, in fact, migration originates among ancestral resident populations, often from lower latitudes similar or equivalent to the current winter range, and that the motive force is not bad weather but increased reproductive success resulting from a bird's ability to capitalize in a fitness sense on seasonal resource flushes by movement. I will have a great deal more to say about this hypothesis, but for the present, it is sufficient to note that, regardless of which theory is right, many birds that breed in the temperate and boreal zones of North America head south in the fall.

Why?

Les Murray's poem "Migratory" provides a superb response, describing the mysterious compulsion of movement as "the wrongness of here." Something inside tells the bird that it is time to go. This is the naked truth. But what?

The short answer is "food." Fall migration is about food. If food remained superabundant, or even simply adequate for survival throughout the coming winter, birds would not migrate. Still, there are some puzzling aspects to the movement, in particular the fact that most migrants leave long before food becomes a problem. This fact is questioned by many of my professional colleagues who believe that the data demonstrate a tight relationship between lack of food and movement. This belief, however, is not supported by data. I made a considerable effort in Chapter 4 of *The Avian Migrant* to provide information on a wide variety of migrants, showing that most leave long before food shortages caused by seasonal change in climate threaten critical supplies, and I will provide exemplary cases over the next several chapters demonstrating clearly that departure timing for fall migration is *not* directly related to food availability.

So if lack of food is not the immediate cause of fall migration for most migrants, what's going on?

There are three factors to consider. The first is physiological: what signals initiate the necessary internal preparations for the journey? The second is behavioral: what tells the bird that this is the moment at which to begin the actual southward flight? The third is evolutionary: why is the bird here in temperate regions to begin with, and why is it leaving?

I define the factors that control timing of the physiological preparations for migration as "distal cues"—that is, they are distant in time by weeks or even months from when southward flight begins. Brilliant lab work with a few species of caged songbirds has determined that one such cue is photoperiod (hours of daylight), at least for long-distance migrants. Each bird apparently has an internal rhythm or biological clock, which controls onset of major physiological processes throughout the annual cycle. The clock appears to be set during the summer months by photoperiod; once set, the clock triggers the appropriate physiological responses necessary for each critical life history event: reproduction, molt, and migration.

A second possible factor is weather, although its importance as a distal cue varies. For long-distance migrants, weather, as a determinant affecting preparations for migration, is minor, keeping in mind that "weather" is quite different from "seasonal change." "Weather" describes the physical, nonbiotic attributes of the environment of a site or region at a given moment. "Climate" refers to these same attributes over a period of time, such as weeks or months.

Departure timing for fall migration for these birds seems to be largely under control of an internal biological clock evolved through natural selection based on seasonal climate change effects on food availability. These birds are referred to as "calendar" or "obligate" migrants since their movement preparations appear to be governed more by date than by external factors. However, weather may act as a distal cue for many short-distance migrants, at least as a modifier of timing dictated by the biological clock. These birds are referred to as "facultative" or "weather" migrants, although even for these species, the biological clock likely plays an important role.

Regardless of the cue, timing of physiological events can be modified by circumstances particular to each bird. For instance, if the reproductive period is prolonged or molt delayed, this may push back the time when preparations for migration will begin.

Nearly all of what we know about timing of migration and the various behavioral and physiological changes associated with the phenomenon is derived from studies of caged birds. German scientist Franz Groebbels pioneered this work, and, indeed, the Germans remain leaders in study of bird migration.

Based on his work, Groebbels described two major physiological/ behavioral states associated with migration, *Zugdisposition* and *Zugstimmung*. Various translations of these terms have been used in the literature since, all of which suffer in my opinion from assumed purpose of the activity, which the original German did not include. German migration scientist Barbara Helm has helped me here. "*Zugdisposition*" can be translated as "preparation for movement," although it is often taken to mean "preparation for migration." I will use the term "feeding state." Similarly, "*Zugstimmung*" means literally "a movement mood" or, perhaps, "actual movement." As in the case of *Zugdisposition*, however, a purpose is usually assigned to the behavior, namely "migratory flight." I will use the term "flying state." As I will discuss further on, research has demonstrated that some species of nonmigratory birds have been found to undergo these activities, presumably as part of dispersal. Therefore, assigning these behaviors solely to migrants is incorrect and, in fact, inhibits our understanding of what the behaviors and their associated physiological changes mean.

Once the bird receives the distal cue or cues, it enters *Zugdisposition* or the "feeding" state—a sequence of actions to prepare for the migratory journey. These actions, both behavioral and physiological, are presumably controlled hormonally, but the actual hormones involved are not well understood.

Behaviorally, a bird in the feeding state eats intensively and, in some cases, may defend an individual feeding territory for this purpose. Physiologically, it lays down subcutaneous fat reserves that can amount to as much as half the bird's weight. Other changes occur, as well, including increase in circulatory system and breast muscle size and decrease in size of reproductive and other organs of lesser importance during long-distance flight. For those songbirds tested in the lab, the feeding state, undertaken in preparation for takeoff on migratory flight, generally lasts six to nine days after which the bird is ready for departure. Obviously, such a time schedule would be much more flexible for free-flying birds, depending on availability of necessary resources and their ability to secure them.

I define the signs that apparently govern actual departure on migratory flight as "proximal cues"—that is, they are more or less immediate in terms of their impact. For fall migration among Nearctic migrants, these cues appear to be related to weather, although the precise nature has not been settled and may in fact vary by species. Factors suggested as cues include direction of winds aloft (north winds in general are favorable for southbound migrants, providing a boost in flight speed), barometric pressure, temperature, humidity, atmospheric stability, and cloud cover. However, there may be behavioral cues as well; that is, when a bird notices departure of other migrants in its vicinity, this may trigger its own departure.

The behavioral/physiological state associated with actual departure on migratory flight is, as described above, called "*Zugstimmung*," or the "flying state." If the bird is a nocturnal migrant, as most songbirds are, the hours during the afternoon and evening prior to departure often are spent quietly resting, which may, in fact, represent a separate physiological state. When the time for departure arrives, the bird voids its digestive tract and launches into the air, often joining loose, mixed flocks of migrants that may include members of its own species as well as other species of songbirds (just as pictured in Charley Harper's wonderful painting *Mystery of the Missing Migrants*). For most individuals, the flight continues through the night until morning, unless dawn finds the traveler over desert, water, or some other inhospitable environ, in which case the flight must continue until a more suitable landing area is reached.

Texas Field Studies of Fall Migration

I was aware of these basic elements of the migratory process when I first began field work on the phenomenon in August 1973, thanks to a fellowship from the Welder Wildlife Foundation. Welder was then, and is now, an extraordinary organization for field biology studies. The foundation was established by a codicil in rancher Rob Welder's will, which required simply that a foundation be set up "to further the education of the people of Texas and elsewhere in wildlife conservation." This idea was translated into reality by the brilliant decision of the trustees to hire Dr. Clarence Cottam as director. Dr. Cottam was one of the preeminent wildlife biologists of his day. As a scientist in the US Fish and Wildlife Service, he was the first (1946) to publish a study on the deleterious effects of DDT on wetlands ecosystems, and as assistant director of that organization, he was a promoter of the work of Rachel Carson, cited by her in the acknowledgments of her book *Silent Spring*. The Welder trustees hired him away from Brigham Young University, where he was dean of the College of Agriculture (and a Bishop of the Mormon Church). It was Cottam who translated Rob Welder's rather vague vision into reality—graduate student heaven.

The Welder Wildlife Refuge is seven square miles of native thorn forest, oak savanna, riparian woodlands, marsh, and pasture located along

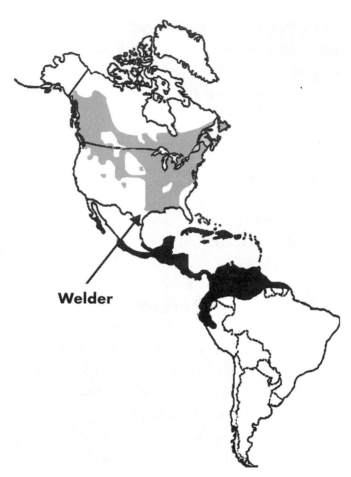

FIG. 3.1. Welder location in relation to the breeding (*gray*) and wintering (*black*) range of the American Redstart.

the south bank of the Aransas River, about ten miles inland from Copano Bay. The campus is composed of several beautiful structures in Mexican colonial style, including a library, a laboratory, a museum, student cubicles, and a lecture hall, as well as buildings for administration and for student and senior staff housing.

Two characteristics make Welder ideal for work on songbird migration. The first is its proximity to the west coast of the Gulf of Mexico, the main migration corridor in the Western Hemisphere. Second is its sub-

FIG. 3.2. Mist net ready to catch birds at Jim Berry's study site at Sunny-side Marsh, Greenhurst, New York.

tropical location, south of the breeding range for most North American migrants but north of the winter range (Figure 3.1), which allowed us to assume that most of the migrants found were birds of passage.

My study site at Welder was a strip of riparian forest, known as Hack-berry Mott—hackberry, cedar elm, pecan, and anacua (a tropical mid-sto-ry tree species) were the dominant trees—about half a mile long and three to four hundred yards wide located along the banks of the Aransas River.

The principal tool used in my work at Welder and elsewhere during much of my career was the mist net. The ones I used were forty feet long, eight feet high, and made of fine black nylon thread. The net is separated along its length by strings of thicker thread into four panels, or shelves. Excess netting is provided so that a loose bag hangs along the lower thread of each shelf (Figure 3.2). When the bird flies into the net, it tumbles into this bag and becomes entangled.

My wife, Bonnie, and I set up fifty mist nets in Hackberry Mott in late summer of 1973, with about one net every hundred feet or so, perpendicular to the lane traversing the mott from north to south.

We spent two field seasons netting birds during the fall at Welder: August to October 1973 and August to October 1974, accumulating 36,000 net-hours (one net open for one hour) in 1973 and 23,400 net-hours in 1974, and capturing 3,350 birds of 96 species. Each captive was aged and sexed (when possible), weighed, measured (wing and tail), checked for molt and amount of subcutaneous fat, banded with a US Fish and Wildlife Service band, and released. Recaptures were reweighed and released.

The main findings were as follows:

1. *Among the earliest transient songbirds captured in our nets at Welder (in August) were birds whose main breeding populations were located in the southeastern United States.* These birds included the Kentucky Warbler, Blue-gray Gnatcatcher, Louisiana Waterthrush, Orchard Oriole, Chuck-will's-widow, Prothonotary Warbler, Black-and-white Warbler, Yellow Warbler, Ruby-throated Hummingbird, Yellow-breasted Chat, and Indigo Bunting.

2. *A second class of early transients were adult flycatchers of the genus* Empidonax. These included the Yellow-bellied Flycatcher, Alder Flycatcher, Willow Flycatcher, Acadian Flycatcher, and Least Flycatcher. Unlike most songbirds, which molt on the breeding grounds, adults of these species molt after migration, on arrival on the wintering ground. Juveniles complete molt before leaving on migration, thus delaying their departure from the breeding grounds by about three weeks, relative to that of adults of the same species. This finding further emphasizes the disconnect in terms of timing of fall migration for members of different age groups, and it further emphasizes the lack of direct influence of weather on timing. These kinds of differences likely occur in all migratory species but are seldom so clear-cut.

3. *Median fall passage dates for the different age groups in species for which we had ample sample sizes were remarkably precise from one*

year to the next. The median passage date for Yellow-bellied Fly-catcher adults occurred on August 22 in both the 1973 and 1974 field seasons. Median passage date for juveniles of this species was twenty-four days later (September 15) in 1973 and twenty-seven days later (September 18) in 1974. This statistic is extraordinary when one considers the vast breeding range from which captures at Welder of these groups are potentially drawn. Given the myriad of likely variations in environment in terms of food availability and weather conditions, one can only see in this remarkable concordance a confirmation of the fact that departure is under genetic control (and not influenced by weather so far as annual timing of departure is concerned).

4. *Yellow-bellied Flycatcher juveniles captured during our two fall field seasons at Welder (427 birds) greatly exceeded the number of adults (90 birds).* The other species for which we had large sample sizes (greater than one hundred captures), such as the Canada Warbler and Mourning Warbler, showed comparable age discrepancy, with juvenile birds significantly exceeding number of adults. Many researchers have reported similar disproportionate representation of juveniles during fall migration, a phenomenon known as the "coastal effect." Such skewed samples are most often found along the borders of obstacles to in-flight rest or refueling, such as oceans, mountain ranges, or deserts.

5. *Most transients captured at the mott stayed for less than one day, regardless of species, arriving at dawn and departing at dusk.* Fewer than five percent of birds captured were recaptured a day or more after capture. Initially (that is, for the first thirty years of my work on migrants), I thought this meant that the birds had sufficient fat resources to continue migration that evening, and therefore had no need to stay additional days. I now believe that they were able to obtain food and rebuild reserves during their brief stay because food was superabundant.

6. *Evidence of territoriality among transients (vocalization, chase, site*

fidelity) was uncommon and limited to only a few species. Members of most species foraged in large, loose mixed-species flocks throughout the day with little evidence of interaction between individuals of the same or different species. This finding indicates that the foods necessary for rebuilding fat reserves for most transients at Welder (that is, invertebrates of various types) were superabundant. As will be discussed later in the text, territoriality is a behavior used when a critical resource, such as food or mates, is defendable and in short supply.

Migration is a behavior that occurs because of superabundant food resources. Regardless of where a migrant's antecedents originated (breeding ground, wintering ground, or someplace else), the migrant individual could not make a long journey, or persist in places along the route, unless food resources were in excess of what the resident avifauna could defend and/or consume. This truism should have been obvious to me during my first weeks in the field in Texas back in fall of 1973. But it was not. I assumed that the thousands of birds, of twenty or thirty transient species, present in Hackberry Mott on any given day already had sufficient resources in the form of fat reserves to continue their journey that evening, as nearly all of them did. The five percent that remained were what confused me. Of these, some individuals of some species set up and defended territories (vocalization, attack), to which they showed site fidelity. I assumed that these birds stayed in order to rebuild fat reserves, an assumption subsequently shown to be correct, at least for Northern Waterthrushes. What I did not understand was that the difference between the ninety-five percent that stayed a single day and the five percent that stayed on for additional days was not necessarily solely a function of fat reserves; it depended on the type, distribution, and number of competitors for the food required. For Northern Waterthrushes, and perhaps some other ground- and understory-foraging species, the foods they needed (ground-dwelling invertebrates) to rebuild fat reserves in my Hackberry Mott study area at Welder during the spring of 1975 were sufficiently dispersed, and the number of potential competitors few enough, to allow economic defense of a feeding territory. Most of the other transient species stopping over at our Texas woodland study site during both fall and spring migration were mid- or upper-story foragers on flying insects and other foliage invertebrates. These food resources apparently occurred in such high density during the migratory periods that establishment of a territory was not necessary, and, in any event, the number of competitors was too large to make defense economical. At the time, I did not believe that prey density could be sufficient to allow so many birds to rebuild fat reserves at the same time. Now, I do.

Thus, I believe that MacArthur and other Northern Home theorists were correct in pointing out that migrants are dependent upon food super-abundances when they leave their breeding areas and indeed throughout their migratory journey. While some individuals of some species do establish temporary feeding territories in transit, the majority appear to move in large, loose flocks and are able to find sufficient food to fuel their journey.

Fall Migration according to the Dispersal Theory

As mentioned earlier, mean departure timing from the breeding (or post-breeding) ground usually differs among the different age and sex categories for most migrant species. The reasons are obvious. Once the young are independent of their parents, each member of the family is on their own lookout, responsible for their own food resources, roosting sites, molt, and migration. In addition, the life history strategies for the various sex and age groups can be quite different. Adult males, for instance, have to devote some time to making sure that they have a breeding territory for the next reproductive season, which is presumably why one hears occasional song in late summer and early fall, long after the young have reached independence. Val Nolan documented this behavior in Prairie Warblers on his Indiana study site, where adult males sang and fought on their breeding territories in September—long after females and young had drifted off.

In some species, such as Song Sparrows, Field Sparrows, and Eastern Bluebirds, males will persist on the breeding territory long after adult females and young birds have headed southward on migration, and they will try to remain on the territory as long as there is enough food to allow them to. Even if they are compelled to leave by lack of food, they will travel much shorter distances and return much earlier in the following spring than females.

Some researchers maintain that this behavior by males is a result of dominance. Adult males push females and young out, forcing them to migrate hundreds or thousands of miles farther than would be the case without male interference.

Perhaps.

But it is also possible that it is females and young that migrate the optimal distance for survival while males must balance survival against probability of having a breeding territory the following year.

Long-distance migrants heading for wintering grounds in the tropics, whether male or female, normally do not have the option of remaining on or near the breeding territory. Their food resources are completely absent from these areas during the winter months. Nevertheless, their departure timing generally occurs weeks or months before any weather-related depletion of resources on the breeding ground occurs. Each individual follows its own schedule in this regard, based on its own environment and experience, although there are large overlaps in both timing and direction taken to the wintering ground. These overlaps result in a mean or average route along which members of a particular age or sex group of a migrant population can be found at highest density at any given moment during the migration period.

If migrants were wanderers pushed south by weather during the winter period to latitudes where superabundant resources allowed them to pass the winter, then we would not expect them to have "routes," which, of course, imply a specific destination. Most members of some species of migrants seem to spend much of the wintering period in this wandering fashion, such as the American Robin. A small percentage of this species migrates all the way to the tropics to winter, but most seem to move in loose flocks into temperate bottomland forest, or even parklands, lawns, and agricultural fields, until heading back north in spring.

Such behavior is not true for most long-distance migrants. Members of these species follow specific routes to particular habitats located in well-defined winter ranges. How and why make up the key mysteries of the migration phenomenon.

We do know something about the "how" part, especially as it relates to navigation. Extensive laboratory experiments have demonstrated that migrants—and residents as well, much orientation work has been done with the nonmigratory domestic pigeon (Rock Pigeon)—are capable of using a variety of environmental cues to determine the direction in which they need to travel to arrive at the desired location. These cues include the sun's location, polarized light, star patterns, the earth's magnetic field, wind direction, and perhaps even infrasound and odors. Still, these capabilities leave large portions of the "how" unexplained. Consider that if you were attempting to travel from your home in Bronxville, New York, to your hotel in Cancún, Mexico, without a map or road signs, you would need to know the latitude and longitude of your starting location and destination and your travel speed; as well, you would need a compass, a chronometer, and a sextant, along with the mathematical tools to know how to use them. Yet six of ten Hooded Warblers on our study site in the Tuxtla Mountains accomplished a comparable feat in traveling from their winter territories at Playa Escondida in southern Veracruz to their breeding territories somewhere in the eastern United States and back again months later. Which means that, in addition to the direction-finding capability demonstrated by returning to southern Mexico, they memorized landmarks so that they could return to the exact same spot where they spent the previous winter.

And that's for adults. The mystery is far deeper for juveniles. They have never been to the winter range. So how do they know how to end up there, which most of them seem to do? Easy, you might say. Just tag along with the adults. Except they don't (bar a few types of birds such as geese and cranes). As I have said, each individual follows its own travel schedule, adult or juvenile, male or female. It is true that many will join loose flocks of conspecifics as they set out on each day's or night's flight, but differences in personal history, such as timing and duration of molt and variations in competitive ability for critical resources, often mean that adults travel

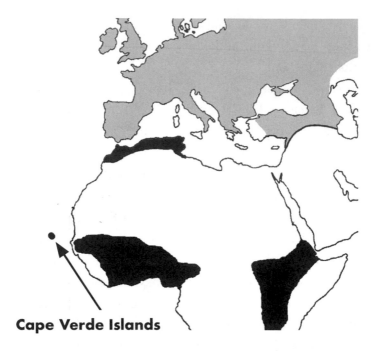

Cape Verde Islands

FIG. 3.3. Breeding (*gray*) and wintering (*black*) range for the Eurasian Blackcap. Birds from the Cape Verde Islands are permanent residents.

on different dates from juveniles, as mentioned above. In addition, as I will discuss below, adults may follow completely different routes from juveniles. As a result, any given migratory flock may be composed largely or entirely of juvenile birds, especially if located along the periphery of a potential obstacle such as an ocean.

Laboratory experiments by Peter Berthold, Eberhard Gwinner, and their students at the famed Max Planck Institute for Ornithology in Radolfzell, Germany, suggest an answer for how such juveniles are able to find where they need to go for the winter without help from adults. The researchers performed a series of experiments involving Eurasian Blackcaps, a small, Old World warbler. The experimental subjects were juvenile birds hatched and hand-reared during the summer in the Radolfzell lab; they were derived from three different populations: (1) a migratory population known to breed in western Europe and winter in western Africa;

FIG. 3.4. Chris Barkan's vision of a surprising experimental result at the Radolfzell lab. DRAWING BY CHRISTOPHER PAUL LYMAN BARKAN

(2) a migratory population known to breed in eastern Europe and winter in eastern Africa; (3) a nonmigratory population from Cape Verde off the northwest African coast (Figure 3.3).

Like most songbirds, blackcaps are nocturnal migrants, and when the time for the initiation of migratory flight (*Zugstimmung*) arrives, they attempt to take off at dusk. If they are prevented in some way, as by being held in a small cage, they will simply repeat takeoff over and over throughout the night, and for each night thereafter for weeks. This frustration of migratory flight is called *Zugunruhe*, which has been translated as "migratory restlessness," although it might be more accurately referred to as "caged" or "frustrated" movement. The researchers placed juveniles into cages where the direction of each takeoff could be measured (Figure 3.4).

Interestingly, birds from the three different populations showed different mean directions for their takeoffs: (1) birds from the eastern European population took off in the direction of their east African winter quarters; (2) birds from the western European population took off in the direction of their west African winter quarters; and (3) the birds from Cape Verde, although nonmigratory, took off in the direction of Cape Verde. Also, duration of restlessness varied by distance to wintering area, with the Cape Verde population showing shortest mean duration by quite a bit (although duration of *Zugunruhe* bore no obvious relationship to the actual time it would take to travel from the German lab to the population's wintering or home area).

Based on these and similar findings from a few other species, Berthold and his colleagues concluded that the route itself was the result of a genetic program, as was duration of frustrated migratory movement. The idea was that following a precisely programmed direction for a precisely programmed time would bring the naïve juvenile bird to the correct wintering area without any need for guidance. This, of course, is true, barring any untoward environmental disruptions, like crosswinds or personal problems, such as not being able to find enough food to fuel long over-water flights.

Further experiments by one of Berthold's students, the late Andreas Helbig, which seemed to clinch the validity of this hypothesis, showed that hybrids between eastern and western populations demonstrated intermediate mean direction and timings, and that repeated selection and breeding of birds that showed outlier orientation could produce changes in *Zugunruhe* direction and duration within a few generations.

The chief problem for the Route Program Theory is its failure to account for the migratory behavior of real birds outside the lab. Here are examples:

1. *Juvenile birds often follow different routes from adults* (as we found at Welder for Canada Warbler, Mourning Warbler, and Yellow-bellied Flycatcher). Migrants tend to take the shortest distance between two points in traveling from breeding to wintering area, but geographical and environmental factors can affect this basic structure significantly, and often quite differently for adults and juveniles. For instance, there is a fall migration pathway called the "North Atlantic route" involving a departure from coastal north-

eastern North America to northeastern South America. Most adults of several migrant species, such as the Blackpoll Warbler, follow this route, but most juveniles of the same species do not. Instead, they follow the eastern coastline of the continent southward to Florida before launching out across the Gulf of Mexico and Caribbean to South America, presumably because they fail to store sufficient fat reserves to fuel the seventy-hour over-water flight required by the North Atlantic route.

2. *Juvenile birds physically displaced from their hatching area orient correctly toward the appropriate wintering ground.* More than eighty studies have been done in which the orientation direction is measured for juveniles that have been moved hundreds of miles from where they were born. In general, these studies find that juveniles orient correctly toward the appropriate wintering area, *not* in the direction that a preprogrammed route to the wintering ground would take them. In other words, they appear to be able to correct navigation direction, changing the route as needed, to migrate to the appropriate wintering area.

3. *Migration can appear and disappear in a matter of a few generations in a particular species.* There are many examples of this occurrence, several of which will be discussed in the chapter on origin of migration (Chapter 9). To cite one, the Bachman's Sparrow was a resident of southeastern US pinelands at the time of European arrival but, with clearing of forest in the 1800s, developed a migratory population that bred as far north as Pennsylvania and Ohio, wintering within the original southeastern range. Evolution of routes supposedly takes thousands of years, but juveniles of this species managed the feat within a generation, obviously without time to evolve a route program.

4. *Correct homing to winter range by juveniles reared in recently expanded breeding ranges.* There are several examples of this behavior as well, but one of the most remarkable is that of the Northern Wheatear. This species, which breeds throughout Eurasia and

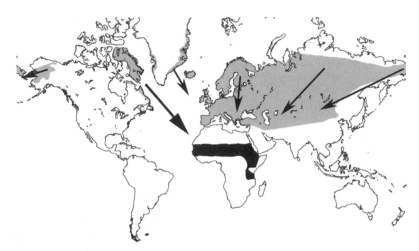

FIG. 3.5. Breeding *(gray)* and wintering *(black)* distributions for the Northern Wheatear. Arrows show southbound migration routes, including those followed by the recently established Alaskan and northeastern Canadian populations.

winters in sub-Saharan Africa, recently (within the past century) expanded its breeding range into northeastern Canada and Alaska (Figure 3.5). If juveniles raised in these new areas were to follow a genetic program for route and timing, they would end up in the middle of the ocean not in central Africa. Yet they manage to orient properly and negotiate the thousands of miles added to the original route to arrive at the African winter range.

5. *Juvenile birds in some migrant species winter in completely different areas than adults, hundreds of miles apart.* Ospreys, Pectoral Sandpipers, and some other long-lived migrant species are known for this behavior. Juveniles follow one route to a wintering area different from adults during their first fall migration, and they follow a different route to a different wintering area during subsequent migrations.

6. *Migratory restlessness and orientation toward area of origin in displaced tropical birds.* This last finding, which comes from the

Radolfzell caged-bird experiments, is perhaps the most intriguing of all of the various lines of evidence contrary to the Route Program Theory, at least for me. As described above, researchers found that juveniles from the Cape Verde Islands population of Eurasian Blackcaps showed migratory restlessness and oriented toward the Cape Verde Islands when tested in a lab in southern Germany, even though these birds were derived from a population that is *resident in the tropics*.

Weather and Fall Migration

Much of the confusion regarding fall migration derives from one critical assumption, namely that birds head south in autumn because of deterioration of the weather. This assumption, however, is no more true than the assumption that they head north in spring because of weather, or that migration originated because of weather. As I have described above, weather may serve as a proximal cue, but it is not the ultimate (evolutionary) cause for migratory movement. That honor goes to food resources. I will discuss this theory further in Chapter 9, "Origin and Evolution." However, I think that the portion of it that applies to fall migration deserves some elucidation here.

Consider the above-mentioned cage experiments, as well as hundreds of other experiments on several different species. Obviously, no weather whatsoever was involved in initiation of the birds' frustrated attempts to migrate. They are, after all, in a cage in a lab.

As I sit at my desk today, at one in the afternoon in my house in Jamestown, New York, on the third of December in the year of our Lord 2018, the weather outside is 38 degrees Fahrenheit with thick cloud cover, drizzle, and a breeze from the north. Last week at this time, there was over a foot and a half of snow on the ground. Today, little of that remains after yesterday's high of 59 degrees Fahrenheit and with rain through the night. Tonight, it is supposed to be in the teens. For my feathered friends of the western New York avian community, none of this matters much. Birds are among the most physiologically capable organisms on earth, able to withstand extraordinary extremes of heat and cold so long as they have the necessary resources (food and shelter). The ones that are still here, the chickadees, nuthatches, woodpeckers, hawks, owls, and so forth, evidently do have the necessary resources, because here they are and here they will stay right through the wintry blasts 'til spring. Most others, indeed more than seventy percent of those that breed here, have left. But it wasn't because of the weather. It was because of resources—specifically, food availability.

"But that amounts to the exact same thing! Weather has direct effects on prey!" you may say.

Now, don't get too excited. Your observation is quite true, but consider that each species has its own resource requirements. Despite the fact that every individual of the 140 or so species that breed in this region was subjected to identical weather throughout the past summer and autumn, each followed its own Tao, depending not only on its species but also on its age, sex, condition, competitive ability, and personal experience.

Take, for instance, an adult male Purple Martin. We'll call him "Joe." Joe spent the winter of 2017–18 in southern Brazil, roosting in a huge flock in downtown Rio at night and foraging out over agricultural areas during the day. He left there in early March and arrived back at his western New York breeding colony at the Boys' and Girls' Club on the Chautauqua Institution grounds on April 23, a week or so before his mate. Together, they raised four young, and by mid-September, Joe had joined thousands of his compatriots on the shores of Lake Erie and had begun the long journey back to his Rio roost. By the end of September, there were zero martins left in New York.

Next, let's consider an adult female Alder Flycatcher named Betty. Her winter territory was located in a patch of scrub on the beach just outside Puerto Cabello, Venezuela. In late May 2018, she got back to the swampy thicket outside Ellery Center in western New York where she had bred in 2017. She dumped Tim, her mate of the previous year, and took up with Bill, a neighbor. Their first clutch was eaten by a chipmunk, but they got one young raised to independence from a second clutch. Of the three others that fledged, two were eaten by a screech owl and one by a kestrel. Betty left for Puerto Cabello on July 20 (Alder Flycatchers essentially have no postbreeding period because they molt on the wintering ground). By the middle of August, the entire North American breeding range was empty of the species.

Then there's Isabel, a female Hooded Merganser. She met her mate, Bob, at Big Lake on the Welder Refuge in South Texas in January 2018. Bob was originally from the St. Croix River along the Minnesota-Wisconsin border, but when it came time to migrate north, he accompanied Isabel back to where she was raised (males normally accompany their mate back to her home for breeding in waterfowl), a beaver pond off Conewango Creek in northwestern Pennsylvania. They departed Texas in early March, arriving at Isabel's beaver pond in mid-April. Isabel immediately began work on lining with breast feathers a cavity fifty feet up in a dead poplar (selected the previous summer). She and Bob copulated, and Isabel began laying her eggs in late April. By early May, the clutch was complete, and Isabel began incubation. With his parental duties completed, Bob left, heading west for Lake Erie, where he would remain until October, undergoing a complete molt, and then heading south for Texas. Isabel's brood of eight hatched in early June. She stayed with them at the pond until they could fly on their own in mid-August, when they all left the pond to spend the next couple of months on the Chautauqua Lake Outlet until migrating south to Texas in mid-October.

These fanciful vignettes are meant to make a point regarding the relationship between weather and food for migrants. The birds whose lives were discussed, as well as all of the other members of their species that bred in this region, experienced equivalent environmental conditions during their time here. Yet the timing of their fall migration was peculiar to each one, as it is, in fact, for every single individual of every migrant species. Why so? This is partly because each species has its own specific needs in terms of food but also because of the peculiarities of distribution and life history unique to each species and each individual of that species.

CHAPTER 4

WINTERING
Period

> Like cranes
> in clamorous lines before the face of heaven,
> beating away from winter's gloom and storms,
> over the streams of Ocean, hoarsely calling,
> to bring a slaughter on the Pygmy warriors.
>
> —Homer, *The Iliad* (Robert Fitzgerald translation)

It is clear from *The Iliad* that the ancients knew something about migration. In it, Homer likens an attack by the Trojans on the Greek host to the southward movement of Common Cranes on their way to wage war in the land of the Pygmies as winter approaches. This neatly summarizes the two central conundrums of fall migration: why birds go south and why they go to a specific place for the winter.

Homer's explanation derives from a story in which Hera, insulted by the Queen of the Pygmies, sends cranes southward each year to slaughter the Pygmies. While this may seem a bit fanciful, it is striking to note that cranes *do* migrate thousands of miles from northern Europe to central Africa, home to tribal groups long known as "Pygmies."

What most fascinates me about the poet's account is that he recognizes that winter storms in Hyperborea provide insufficient motivation for a trip to central Africa. Therefore, he falls back on a standby hypothesis: the capriciousness of the gods. While not totally rejecting this explanation, I present alternatives in this chapter for the forces that motivate migratory movement to winter quarters. I also discuss what we know about avian life history during the wintering period based on my own investigations and those of many others, as reported in the literature.

Homer's crane metaphor highlights a key problem for the Northern Home Theory, namely that the winter range of migrants requires further explanation than bad weather. Indeed, a curious aspect of the winter range is that it appears to function as a lodestone for many species of migrants, drawing back to it not only experienced adults but young birds hatched thousands of miles away just months earlier.

Many people, if they have thought about it at all, assume that the young are led to appropriate wintering areas by adults. This assumption is true for a tiny percentage of migrant species, such as some geese, which travel between breeding and wintering areas as family groups. In most other migrants, including all songbirds so far as we know, each individual bird—male or female, adult or juvenile—follows its own calendar of major life history events, such as molt, migrations, and reproduction, shaped to its own specific experience and circumstances. In fact, as noted in the previous chapter, mean timing of migration departure often differs significantly for the various age and sex categories of a species. So how does each individual find its way to the appropriate winter range?

For some species, as mentioned earlier, just heading south until an appropriate habitat is reached seems pretty straightforward. But for birds like the Barred Warbler, with a vast breeding range and miniscule winter range, this approach obviously won't work (Figure 4.1).

So. If not weather or Pygmies, then what, exactly, is the reason for southward migration? The answer is simple. Food. Superabundance of food resulting from seasonal change in spring is what brought the birds north, where they could raise more offspring than would have been the case had they stayed at home. Declines in food availability resulting from changes in climate send them south in fall.

Now, you may say, "But aren't bad weather and disappearance of food

FIG. 4.1. Barred Warbler breeding (*gray*) and wintering (*black*) distribution.

essentially the same thing?" The answer is that they can be, as in the case of a cold snap turning water to ice for waterbirds dependent on aquatic organisms for their livelihood. However, this particular example simply clouds the picture. Hundreds of species migrate despite little or no change in weather.

In order to begin to understand migratory movement, we need to recall the assumed motivating factor for any individual behavior, namely to increase fitness. "Fitness" is the individual's genetic contribution (offspring) to the next breeding generation. Of course, winter is not the time when most birds make that contribution, but they have to survive in order to be able to contribute during the next reproductive season. In other words, maximizing fitness in winter is focused mainly on survival, and the chief element required to maximize winter survival is food. For chickadees, nuthatches, woodpeckers, and the like, food is normally available throughout the north temperate zone winter months, which is why most do not migrate. For tropical species like wood warblers, tanagers, vireos, hummingbirds, and so forth, preferred foods are not available in the higher latitudes, or at least not in sufficient quantity to allow a high probability of survival. So they head south.

But why do they go where they go to spend the winter? Why does a Barred Warbler raised in Switzerland go to the same small region in central Africa to spend the winter as one raised in western Mongolia? The key to understanding this is the concept of probability. The bird winters where its chances of survival are greatest. Now you might think, won't any nice warm place do? Evidently not, because many species have their own winter range, and within that range, their own preferred habitat, outside of which they are rare or absent. Circumscribed winter ranges for migrants pose a serious enigma for the Northern Home Theory. If, as suggested by this theory, birds evolved a migratory habit simply as a result of being pushed south for millions of years by inclement weather into warmer climes where food was not a limiting factor (hence no need for a specific niche in a specific habitat), why should a species have its own unique winter distribution? Shouldn't migratory birds winter just about anywhere in lower latitudes in appropriate habitat? Following the fitness argument, a Northern Home theorist would have to explain small winter ranges located thousands of miles away from the breeding range by assuming that millennia of natural selection favored (in a fitness sense) those individuals that wintered in a particular area over those that did not. But this reasoning is completely at odds with the assumption that food is not limiting during the winter season for migrants. If food is not limiting, then it is not a factor in natural selection. However, if food is limiting, then you (the individual migrant) will have to compete with others for it—and not only with others of the same species but with members of other species that live in these regions throughout the year—and if you are going to be able to compete with those resident birds, then you are going to have to have a resource space (niche) that is entirely your own in that wintering habitat or you won't be able to compete and win. You will starve.

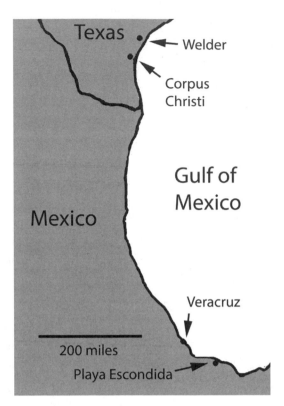

FIG. 4.2. Relative location of the Texas (Welder) and Veracruz (Playa Escondida) study sites.

Field Studies of Wintering Migrants in Veracruz Rain Forest

The arguments presented above formed the basis for my initial studies of migrant winter ecology. As in the case of my work on transient ecology at Welder in South Texas, the funds were provided by the Welder Wildlife Foundation. The main study area for our research on the wintering ecology and behavior of migrants during my thesis work in the winters of 1973–74 and 1974–75 was done on a piece of property known as Playa Escondida,

FIG. 4.3. Bonnie Rappole in the middle of the Playa rain forest study site, November 1973.

which belonged to Raul and Julietta Garcia. This site was located along the southwestern coast of the Gulf of Mexico in the Tuxtla Mountains of southern Veracruz (Figure 4.2).

Vegetation on the site consisted of about ten acres of relatively undisturbed rain forest (Figure 4.3), two acres of rain forest in which the understory was pretty heavily grazed by cattle, and three or four acres of mixed pasture and low, scrubby second growth. As in Texas, our principal tool (after binoculars, of course) was the mist net, although on this site we placed the nets in a grid, rather than a line, with each net within 150 feet of its four neighbors (for interior nets). During the first season (November 1973 to March 1974), we netted almost entirely in lowland rain forest (referred to as *selva* in Spanish); during our second season (November 1974 to March 1975), we netted in both *selva* and the neighboring pasture and overgrown fields.

Captured birds were handled similarly to our Texas birds, so far as data collection and banding was concerned, except that each migrant captured was also given a unique color-band sequence for individual identification in the field prior to release, enabling me to document by observation what birds lived where.

Using these methods, supplemented by hundreds of hours of observation, we were able to learn a great deal that was previously unknown concerning migrants passing the winter in their tropical homes, as summarized below.

1. *Individuals of twenty species of migrants spent the entire winter season, from the time of their arrival in October and November until their departure in March and April, as members of the rain forest avian community.*

2. *Members of the migrant species found in rain forest fell into two broad categories in terms of their behavior and distribution: solitary birds that could usually be seen in the same piece of forest every day, all day, and birds that associated with the mixed-species flocks of resident birds that roamed at mid-canopy level across several acres of forest.* Migrants in the first category were found mostly in the understory and included the Hooded Warbler, Wood Thrush, Kentucky Warbler, Ovenbird, Louisiana Waterthrush (restricted to watercourses), Yellow-bellied Flycatcher, and Gray Catbird (thickets). Migrants found with mixed-species flocks included the Blue-gray Gnatcatcher, White-eyed Vireo, Worm-eating Warbler, Black-and-white Warbler, Magnolia Warbler, American Redstart, Black-throated Green Warbler, and Wilson's Warbler.

3. *Many individual migrants of several of the above-mentioned species spent the entire wintering period on or near the same small piece of forest, as documented by recapture and observation of color-banded birds.*

4. *Individuals of nearly all of these species were heard to vocalize regularly, that is, every hour or so, throughout the day, with peaks in midmorning and early evening; most used chips or chirrs for this purpose, although song also was used occasionally by the Black-and-white Warbler, White-eyed Vireo, and Yellow-bellied Flycatcher.* We were able to document for the vireo that both males and females sang.

5. *Many individuals of some wintering migrant species reacted aggressively to playback of vocalizations.* Using a Nagra tape recorder with parabolic microphone, I recorded examples of most of the vocalizations given by migrants. I used these recordings in playback experiments to see if I could get a reaction (approach to within a few feet of the recorder), like what one might expect from a territorial male bird on the breeding ground. There was quite a range of experimental result depending on species. Hooded Warblers responded rapidly by approaching and obviously searching for the source of the sound more than half of the time; other species, like the Wood Thrush and Kentucky Warbler, responded some of the time, while still others did not respond at all.

6. *The combination of within- and between-season site fidelity and periodic vocalization daily throughout the winter season was indicative of territoriality.* Even those birds, like the Magnolia Warbler and American Redstart, that were seen mostly in mixed-species flocks were usually represented by a single member of their species in these flocks, and they appeared to defend against membership in the flock by additional conspecifics (members of the same species).

7. *Apparent chases between conspecifics were seen on occasion, especially early in the winter season, shortly after arrival.* Such obvious territorial encounters between free-flying individuals were observed in Wood Thrush, Kentucky Warbler, Hooded Warbler, Wilson's Warbler, Yellow-bellied Flycatcher, and Ovenbird.

These initial observations were highly suggestive that migrant life during the winter was not as predicted by the Northern Home Theory. Nevertheless, such sightings amount to little more than suggestive anecdotes, and I was anxious to see if I could produce clear-cut territorial responses on demand. To this end, I developed two experimental procedures for presenting a potential "intruder" onto a suspected territory. The first technique involved tying a bird by a thread to a nail on a board and placing the board in a part of the microhabitat of the suspected territory of a color-banded bird.

The second method used for testing territorial defense involved construction of a small cage, a cubic foot in size, built of sticks and mist net with a wooden floor, into which a bird captured elsewhere would be placed. Then I would put the cage onto what I suspected might be the territory of a bird of the same species that had been previously captured and color-banded.

When I first began these experiments, I would just walk out in the woods and choose a likely place near where I had seen a vocalizing member of the target species. However, except in the case of Hooded Warblers, I found that my presence inhibited response for most species. Therefore, I built a portable blind made of burlap so that my presence would be less obtrusive. From late December until late February, I spent about 150 hours in that blind with camera ready.

Thank goodness for Hooded Warblers. If you stake out a Hooded Warbler on the territory of another Hooded, the owner will respond immediately. It will approach, give stylized displays that are the same for every territorial individual tested, male or female, and, ultimately, attack when the "intruder" doesn't leave. Furthermore, the owner doesn't tire. As long as you leave the "intruder" in place, it will continue to vocalize, display, and attack. Territorial individuals of other species were much more picky in terms of their response to these and similar experiments. Something about the setup seemed to deter them. However, with persistence, I was able to observe and photograph territorial response, including attack and stylized displays, in the Wood Thrush, Kentucky Warbler, Least Flycatcher (in second growth, not forest), Magnolia Warbler, and Worm-eating Warbler.

Wintering migrants on our rain forest study site used specific parts of this habitat, indicative of specific niche requirements. For instance, Louisiana Waterthrushes were found only along watercourses and Gray Catbirds only in thickets. In Raul's *selva*, there were really only two species of migrants that had fairly uniform distribution, with territories throughout his forest: the Wood Thrush and the Hooded Warbler. However, these species demonstrated characteristic usage patterns: Wood Thrushes foraged almost entirely on the forest floor, while Hooded Warblers sallied for flying insects in relatively open parts of the understory a few feet off the ground.

8. *Territories defended by migrants wintering in rain forest had specific boundaries.* Owners spent most of their time during the winter within these boundaries, and they returned to the same piece of ground the following winter. Wood Thrushes were furtive and did not always respond well to caged intruders or playback, but the Hooded Warbler proved to be quite dependable. If you walked out to where you thought Yellow Left (an individual with a color band on his left leg) was supposed to be, you could nearly always find him (most Hoodeds with territories in *selva* were hims), or, if not, you could bring him in with playback. Also, Hoodeds are not furtive. As noted above, they forage mostly three to six feet above the ground by sallying for flying insects, and they don't change their behavior much when approached by an observer. These characteristics made me wonder if I might not be able to plot the boundaries of Hooded Warbler territories. We did not have GPS back then, so all of the location points for individuals had to be marked down on a hand-drawn map made from data collected with a compass, a one-hundred-foot measuring tape, and pink flagging tape. Using these items, I set out to map the territory boundaries for ten Hooded Warblers I knew from banding and observational data to be territory owners in Raul's *selva*. I used three methods to obtain points: recaptures and resightings; response to playback and stakeouts; and use of the "territory flush" technique developed by John Wiens to determine boundaries for grassland bird breeding territories (also described by David Lack for determination of territory boundaries for the European Robin). Most of the data, however, came from the flush technique, the others being far too time-consuming. This method works for Hoodeds because of their visibility, uniform distribution, and relative tameness. It involves two "beaters" slowly walking through the understory "pushing" the territory owner toward his or her boundary. When he or she gets there, or a little

ways beyond it, the bird will fly back over or around the beaters, often after being attacked by a neighbor. The point the bird flew from is marked with tape and on the map. We repeated the process until the boundaries for all ten birds had been mapped. Similar procedures for testing of territoriality were followed during our second winter of work in the rain forest and in nearby pasture and second growth at Playa Escondida (November 1974 to April 1975). The most striking discovery of this second season was the rate of return of migrants banded the previous year. Other researchers, such as Paul Schwartz (for Northern Waterthrushes wintering in Venezuela), had reported returns of banded North American migrants to tropical sites where they had wintered previously. However, no other researchers had mapped the territory boundaries for each individual in a population prior to departure in order to test the precision and rate of return for the next winter season. As described above, we had done this for the Hooded Warbler, mapping ten rain forest territories during the 1973–1974 season. When we arrived in December to commence the 1974–1975 season, one of our first activities was to capture (or recapture) all of the Hooded Warblers on the site and map the boundaries of their territories. This process took us about a month of intensive mist-netting, playback of territorial chips, and "pushing" using the flush technique in order to get all of the resident territory owners color-banded and to find out exactly where the boundaries of their territories were located. The results were stunning. Six out of the ten birds that had held territories on the site in 1973–1974 had returned to reoccupy those same territories in 1974–1975, and the boundaries of those territories were roughly the same as they had been the previous year.

This phenomenon was not restricted to Hooded Warblers. We had returns for all of the common migrant species of our rain forest community, but the return rates for these species were far lower than for Hooded Warblers for two main reasons. First, we had made no special effort to capture and band every territorial individual or map their territory boundaries during the 1973–1974 season for species other than Hoodeds. Second, during the 1974–1974 season, we made no special effort to find all birds banded during the previous year for any species other than Hoodeds. Had we made comparable efforts for the other territorial migrants wintering on our *selva* study plot, I have no doubt that we would have found comparable rates of return to those found for the Hooded Warbler.

9. *Using these methods, we were able to confirm that there was a community of North American migrant species occupying and showing both short-term (within a given winter season) and long-term (from one year to the next) site fidelity in the quintessential primary habitat of the tropics—rain forest.* This migrant community in the lowland *selva* of the Tuxtlas consisted of Ovenbird and Kentucky Warbler at ground level; Louisiana Waterthrush along streams; Northern Waterthrush around swales and ponds; Gray Catbird, White-eyed Vireo, and, occasionally, Least Flycatcher in thickets and tree-falls; Hooded Warbler, Wood Thrush, and Yellow-bellied Flycatcher in the understory; Magnolia Warbler, Wilson's Warbler, Worm-eating Warbler, Summer Tanager, Black-and-white Warbler, Blue-gray Gnatcatcher, and American Redstart in the mid- and upper story usually (and perhaps obligate) as members of mixed-species flocks. Black-throated Green Warbler, Black-throated Blue Warbler, and Swainson's Thrush were also found on our *selva* site several times in both seasons, and we had individual records for Broad-winged Hawk and Ruby-throated Hummingbird.

10. *Members of several species of migrants spending the winter in rain forest defended individual territories, using vocalizations, visual displays, and direct attacks against potential conspecific intruders. Individuals of nearly all of the species of migrants that wintered commonly in rain forest habitat showed some evidence of territorial defense by members of both sexes including regular (that is, hourly) use of vocalization (song or call) as advertisement, response (approach) to playback of vocalizations, use of stylized display, attack, and chase of "intruders" (free-flying, caged, tethered, or mounts) of the same species.* These evidences of defense were most marked during the period shortly after fall arrival on winter quarters, but regular vocalization persisted throughout the entire winter season until birds departed for the breeding ground. Strength of response to intruders varied markedly among species, and even from individual to individual of the same species. Hooded Warblers, for instance, could be depended upon to approach in reaction to initial playback or first-

time introduction of a caged bird or mount about half of the time, and once "sensitized" (that is after finding an "intruder"), almost always approached, displayed, and attacked on subsequent exposure. Ovenbirds and Kentucky Warblers were far less aggressive to artificial intruders (caged birds or mounts), although they responded immediately to free-flying interlopers. I did more work with the Hooded Warbler than with any other migrant found wintering in *selva* because of its abundance, visibility, and ease of capture. Visual display patterns used in defense of territory are probably no more varied in this species than in any of the others known to defend winter territories in the Tuxtlas, but Hoodeds responded more readily to artificially introduced trespassers, so more complete descriptions and films of displays could be obtained. Displays recorded included the following: (1) "Wing droop"—this pattern is a basic one from which other displays are formed. The wings are held slightly out, the head is pulled in and the feathers fluffed so that the bird has a hunched appearance. The bird often assumes this position when landing near an intruder. It will then hop down the branch toward the intruder, maintaining this attitude. (2) "Head switch"—this display is a variation of the "wing droop." The bird's body is turned away from the intruder, and the bird turns its head from side to side. (3) "Upward"—in high-intensity situations, the bird faces the intruder, standing quite close to it (six inches or so), and moves up and down from this position to the "wing droop" (Figure 4.4).

Direct attacks and bill snapping are also included in the display repertoire. Displays seem to exploit the gold cheek patch framed by the black hood and bib. Fewer than five percent of females possess this pattern, yet they appear perfectly capable of defending territory against male intruders. To test the relative importance of vocalization, color patterns, and movement in stimulating territory-owner response to an intruder, forty-five separate experiments were performed in which different combinations of a live bird, voice, and a stuffed bird were placed in the territory of a marked bird. The results present a fairly clear pattern of the aggressive territorial response to intruders. An intruder sitting still is unlikely to be seen let alone attacked. If it is seen, it will be approached and investigated, possibly displayed to, and possibly attacked. A vocalization is usually investigated

FIG. 4.4. Displays used by both male and female Hooded Warblers in defense of winter territory: "wing droop," "head switch," and "upward." DRAWINGS BY CHRISTOPHER PAUL LYMAN BARKAN

and, if a bird of the "right" color pattern is found, it is usually displayed to and attacked, often repeatedly. On a few occasions, sessions went more than three hours with a territorial bird attacking a caged bird or stuffed mount at approximately one-minute intervals throughout that time. Once a territory owner had seen a mount or caged bird and heard an accompanying vocalization, it often would continue attacking and displaying whether the intruder stopped vocalizing or not.

I performed "caged intruder" and stakeout experiments, like those described above for the Hooded Warbler, on nine other migrant species. I found that color patterns, sound, and behavior were important factors in triggering a stylized, aggressive response for all species studied in this way, although the relative importance of these factors varied considerably from species to species. The Hooded and Magnolia warblers showed a high degree of sensitivity to the species-specific voice (call note) and color pattern. In other species, such as the Wood Thrush, Kentucky Warbler, and Ovenbird, behavior seemed more important than color pattern in eliciting response. Experiments with the latter group of species followed a rather frustrating pattern. The territorial bird would approach in response to the playback, inspect the caged intruder or stuffed bird, occasionally give low-intensity displays, and then leave. Only on the rare occasions in which the caged bird behaved in exactly the "right" way was a higher-intensity response obtained. This behavior was in direct contrast to the situations involving free-flying birds of these same species, where attack was immediate. These situations are of little use in determining visual displays because elaborate displays are not usually used on intruders; these displays are seen after fall arrival during territory establishment or in occasional boundary disputes with neighbors later in the season.

11. *Death rates for territorial individuals of migrant species wintering in rain forest were near zero.* Based on observation of color-banded birds and recaptures over the course of the wintering period, none of the territorial birds died of natural causes during either the 1973–1974 or 1974–1975 field seasons. Many find this assertion unbelievable given the variety of threats in a rain forest environment. Nevertheless, it is true. I suggest that the reason for the survival success of birds on their winter territories can be attributed to knowledge of their surroundings. Based on our radio-tracking work and observations, there are really only two major predators on forest songbirds in the Tuxtlas: Ferruginous Pygmy-Owls and Barred Forest-Falcons. Birds sitting on their territories throughout the winter know the habits and movements of their local predators very well. Wanderers don't, which makes them considerably more vulnerable to being somebody's dinner.

In February 1975, Raul had all of the forest south of the road cleared as pasture. Of the four Hooded Warblers with territories in that area, two were later recaptured as floaters in the area north of the road; the others simply disappeared.

Territory holders spent most of their time actually on their territory during the wintering period. Once territory was established on arrival in the fall (or on replacement of a removed territory holder), the owner showed a high degree of sedentariness, seldom moving beyond its boundaries throughout the winter season. In over seven thousand net-hours on our *selva* site, only twice were territorial birds captured more than 150 feet from their original point of capture, with the exception of the two previously mentioned birds whose territories were destroyed. Tests with caged intruders and mounts showed that winter territories likely were defended continuously from time of establishment in the fall until time of departure in spring.

12. *Wanderers (evidently nonterritorial birds) were observed entering defended areas of known territory holders as late as March.* These birds would replace the territory owner and assume defense if given the opportunity. Long hours spent in the blind revealed some notable insights into the curious functioning of the system, mostly from work with Hooded Warbler territory holders but also from observations of Wood Thrushes, Ovenbirds, and Magnolia and Kentucky warblers as well. Wintering-ground territories were not exclusive. Wanderers would be seen foraging occasionally right through the space defended by a known (color-marked) territory holder. The reaction by the territory holder was fascinating. So long as the intruder made no vocalization and stayed only for a short period, it was ignored, but a chip note or some change in posture by the invader would bring immediate attack and chase. I had noticed a similar response to artificially introduced (caged or tethered) intruders as well. So long as they were quiet and still, they were generally ignored by the owner. Only when they chipped, or I used playback of a chip to accompany their "intrusion" did the owner pay any attention. The origins of free-flying intruders were unknown for the most part. However, in two cases the intruders were birds that had been captured (and color-marked) on neighboring second-growth sites. The same patterns of interaction were directed toward free-flying intruders as toward caged birds. On March 10, 1975, I was in a blind on a territory belonging to a male Hooded Warbler banded with a red color band on the left leg and a green color band on the right leg (referred to as "red-left green-right," or RlGr in shorthand notation). At a little after eleven in the morning, a female Hooded Warbler banded with both a yellow and a white band on the left leg ("yellow white left," or YWl) was seen on RlGr's territory. She had been captured originally about a mile and a half away in shrubby second growth several days previously and not been seen or captured at that site since, indicating that she was a wanderer or floater (nonterritorial bird). When first observed, she was foraging silently and RlGr was not in sight. At about noon, a second intruder onto RlGr's territory was seen, an unbanded male who was captured

later that day and marked with both a yellow band and a white band on the right leg, yellow white right (YWr). Both floaters (YWl and YWr) foraged in front of the blind quietly, often within six feet of each other, with no apparent interaction. When the owner, RlGr, appeared, he chased both floaters out of sight. However, they both came back as soon as he had moved to another part of his territory. This sequence continued throughout the afternoon—the floaters being evidently unwilling to abandon a rich food source (insects hovering around a sewage drain from one of the hotel units). Attacks on them were of two types. In low-intensity situations, a supplanting attack was used in which RlGr fluttered slowly in the direction of the intruder. No fighting or chasing was involved. The second type was given if one of the intruders happened to vocalize. About once an hour, the intruder male (YWr) vocalized, giving the chip note softly. Whenever this calling occurred, RlGr flew from wherever he was in his territory and attacked with a high intensity, rapid rush. Bill snapping accompanied this type of attack. If the intruder saw the attack coming, he or she fled and was chased. If not, the intruder was actually hit and then chased. At about three in the afternoon, I removed the territory owner, RlGr. The two floaters continued to forage peaceably within sight of one another for the rest of the day. The next day (March 11, 1975) both floaters were again seen foraging on RlGr's territory, and at a little after noon, the female (YWl) attacked and chased an unbanded intruder male from the area. At about five thirty in the late afternoon, the banded floater male (YWr) attacked the floater female (YWl) and drove her out, taking over the territory twenty-six hours after the original owner, RlGr, had been removed. I removed YWr at that time. When the area was checked the next day (March 12), the female (YWl) was found to be in sole possession of the territory. In three other cases where known territorial Hooded Warblers were removed from their territories, they were replaced by birds that had not previously been captured. Once established on their new territory, they can become both sedentary and aggressive toward conspecific intruders, whether artificially introduced or free-flying.

13. *For some wintering migrant species, there was no evidence of territorial behavior in birds seen in habitats other than rain forest.* Work in the

overgrown pasture and second-growth habitats yielded another interesting finding. Individuals of several of the same migrant species found in forest were also found in these habitats, where their behavior often was quite different from their *selva*-inhabiting conspecifics. Banding data revealed few recaptures, and there was little evidence of territorial defense. In other words, many of these birds seemed to be wanderers, behaving just as predicted by the theories of MacArthur and his associates.

Whether or not tropical forest second growth is suitable for long-term territory occupancy will, of course, vary by species. Richard Chandler points out that Golden-winged Warblers in Costa Rica remained on, and defended territories in, forest second growth throughout the wintering period.

Sex ratio in the various habitats in which migrants were found wintering varied markedly. For instance, in the Hooded Warbler, of twelve birds known to hold territory in *selva* in 1975, ten were males. Of two birds holding territories in older second growth, both were females. In general, males captured in rain forest outnumbered females eight to one, while females slightly outnumbered males in second growth. These ratios are based on specimen data as well as observations, so that the skew is not caused by females with male-like plumage (which, in any case, compose less than five percent of the female population of Hooded Warblers according to specimens in the Minnesota collection). However, during spring migration in both Veracruz and Texas, the sexes were captured in approximately equal numbers. The Hooded Warbler was not the only species in which sex ratio was found to be skewed. Males outnumbered females captured in *selva* on our study sites in Veracruz, whereas females outnumbered males in second growth, for the Wood Thrush, Wilson's Warbler, and Kentucky Warbler, while females outnumbered males in forest for the Black-and-white Warbler.

14. *There was no evidence of "ecological counterparts" for migrants among the tropical resident species with whom the migrants shared the various wintering habitats of the Tuxtlas, including rain forest.* As explained in Chapter 1, MacArthur predicted that indigenous resident bird species in the tropics would fill the niches occupied by migrants on their temperate breeding grounds, preventing migrants from long-term residency in most major trop-

ical habitats. Therefore, an important part of my field work in both Texas and Mexico was to spend many hours each day observing individuals of the various species of migrants that we found living in the principal habitats of our study areas to see if there was evidence of members of resident indigenous species similar in foraging habits to migrants, and/or aggressive toward them. As luck would have it, the textbook example of the ecological counterpart involved the Hooded Warbler, star performer on my Playa study sites. MacArthur acolyte E. O. Willis had published an important paper relevant to the "ecological counterpart" issue in 1966 in the journal *The Living Bird* in which he described interactions between resident Spotted Antbirds and wintering migrant Hooded Warblers in Belize rain forest. The antbirds, as indicated by their name, appear dependent on colonies of army ants to scare up their food for them. Antbird pairs, male and female, defend a space around the ant colony as it moves through the forest, snatching invertebrates up as they attempt to escape the swarm. Willis observed that Hooded Warblers occasionally attended such swarms as well, from which they were often driven off by antbirds. Based on this observation, he concluded that the Spotted Antbird likely was a tropical forest resident ecological counterpart of the Hooded Warbler. The case collapses in the Tuxtlas, despite the presence of the same army ant species as that found in Belize (*Eciton burchellii*) and an antbird, the Black-faced Antthrush (but no Spotted Antbirds). As discussed above, I found that Hooded Warblers resided on small individual territories in forest throughout the winter period. Certainly, they took advantage of the prey dislodged by the ants when swarms crossed into their territories, but I saw no interactions between Hoodeds and any tropical resident species during such times, let alone the antthrush, which was seldom seen. The closest thing to an indigenous tropical forest counterpart for the migrant Hooded on my study sites was the Sulphur-rumped Flycatcher, a bird comparable in size and foraging behavior but different in microhabitat use (thickets versus open understory).

Similar kinds of analyses were performed for all other migrant members of the Tuxtla *selva* community, and they too were found to show short-term (seasonal) and long-term (annual) site fidelity in tropical rain forest. No evidence for presence of indigenous "ecological counterparts" for migrants was found. In fact, detailed notes on foraging behavior revealed that species most comparable were other migrants, as in the Wilson's Warbler and Magnolia Warbler, which forage at roughly the same height on average in the forest canopy (about sixty feet) and the Ovenbird and Kentucky Warbler, which forage on the forest floor. In both cases, although foraging heights were comparable, foraging strategies were found to be quite different. The Magnolia Warbler forages mostly as a gleaner, picking invertebrates off from the surfaces of leaves, while the Wilson's Warbler forages mostly as a flycatcher, making short sallies after flying insects. Ovenbirds walk along on the forest floor picking invertebrates from the ground and duff, while Kentucky Warblers hop, reaching and jumping upward to glean invertebrates from overhanging leaves. Many hours of observation revealed no aggressive interaction between members of these species pairs.

I have described documenting, by banding and observation of color-banded birds, the presence of members of twenty species of migratory birds on my *selva* study sites throughout the entire winter season; another twenty species or so of migrants were present in neighboring sections of other major habitats, such as wetlands, pasture, and both younger and older second growth. I watched carefully for evidence (similar foraging behavior or active display and attack) of tropical resident species that might qualify as ecological counterparts for the migrants wintering in these environments, and I found none. Of course, the acid test for presence of ecological counterparts for migrants would be migrant absence as long-term residents (weeks or months) from tropical habitats, which the netting, banding, and observations over two field seasons demonstrated was not the case.

CONCLUSION

The import of all of these observations and findings was quite clear, even after just a few months—namely, that many members of the migratory bird species wintering in the forests and fields around Playa Escondida were not wandering interlopers. For the period of their stay in the Tuxtlas, they were "card-carrying" members of the rain forest avian community.

Field Studies of Wintering Migrants along the Texas Gulf Coast

Our project was designed to collect information on the ecology and behavior of North American migrants during the various parts of their life cycle in order to evaluate current understanding of the phenomenon of migration. As such, our work at Welder was focused mainly on migrants in transit. However, South Texas is an important wintering area for many Nearctic species, and we took advantage of that fact to collect what information we could on these during our fall and spring seasons at the site. Our main findings are presented below.

1. *Members of six Nearctic migrant species were common winter residents in our riparian forest study site (Hackberry Mott) and neighboring habitats: Eastern Phoebe, House Wren, Winter Wren, Brown Thrasher, Hermit Thrush, and Lincoln's Sparrow.* Earliest arrivals among these species came in late September, with most arriving by mid-October. Weather, other than winds aloft, had no apparent effect on timing of departure for these species from their breeding areas.

2. *Recapture of banded birds documented that at least some members of these species remained throughout the winter (until departure on spring migration) on the same piece of ground in the mott to which they returned in the fall. These banding data document that wintering members of these species were not wanderers.*

3. *Behavior of these birds, that is, intolerance of conspecifics as expressed through vocalizations and chases, indicated that individual members of both sexes defended exclusive winter territories.* House Wrens of both sexes defended individual winter territories. There was a House Wren living in the bushy plantings around the Welder Study (a large room with cubicles attached to the main administration building by a portico). I captured and banded the bird, releasing it back into its home bushes along the study wall just outside my cubicle window. I then brought a House Wren that I had captured in Hackberry Mott and tied it to a nail on a board, which I placed on top of one of the bushes outside the study; essentially a literal "stake out." It did not take long for the resident, banded House Wren to discover the

FIG. 4.5. Brown Thrasher (*left*), with "ecological counterpart" Long-billed Thrasher (*right*). PHOTOS (LEFT TO RIGHT) BY LAURIE MCCARTY AND KRISTY BAKER

"intruder," which it attacked immediately and continuously for a minute or two until I ended the experiment by intervening and releasing the doubly abused captive. I filmed this event with an 8 mm camera, and although the quality was terrible (you could barely make out the two principals), at least I had documentation.

4. *Eastern Phoebes of both sexes defended individual winter territories.* Phoebes, which breed across much of the eastern United States, occur only as winter residents at Welder, where they begin arriving in mid-October, establishing small territories in woodlands and thorn forest bordering creeks and roads.

5. *For at least two species, the House Wren and Eastern Phoebe, vocalizations used in defense of winter territory included occasional use of the song used by males in defense of territory on the breeding ground. In these species, both males and females sang.*

6. *Potential "ecological counterparts" (sensu MacArthur; see Chapter 1) were present for four species of wintering migrants: the Long-billed Thrasher for the Brown Thrasher (Figure 4.5); the Bewick's Wren for the House Wren; the Carolina Wren for the Winter Wren; and the Olive Sparrow for the Lincoln's Sparrow. However, there was no evidence of competition, either direct or indirect, between members of any of these species pairs.*

7. Departure of wintering birds in spring was unrelated to any particular weather pattern that we could discern.

The Migrant Wintering Period according to the Dispersal Theory

The Dispersal Theory for the origin of migration does not suffer from the problem of attempting to explain restricted winter ranges. The reason migrants go to these places to spend the winter is that it is where they originated from. It is where they evolved. It is the one place where a niche space exists that is entirely their own—the one place where they can outcompete members of any other species for food during lean times.

According to the Northern Home Theory, when they head south to their wintering grounds, migrants will meet their most important competitors for food in the form of the "ecological counterpart," a hypothetical species that is resident in the most stable habitats of lower latitudes. Members of this tropical doppelgänger species occupy a niche similar to that of the potential migrant invader, only they are better at it and able to outcompete the migrant, thereby excluding the migrant from such desirable habitats. Thus, the ecological counterparts for migrants among resident species function as a sort of MacGuffin—a plot device required to help explain why migrants are not found in primary (virgin) habitats like rain forest and cloud forest in the tropics and elsewhere. MacArthur recognized the necessity of ecological counterparts from a theoretical perspective, and a few field biologists looked hard at wintering-period migrant-resident interactions for evidence of the phenomenon. Some researchers felt that they had found data supporting the concept in the chases that occurred at resource concentrations in tropical wintering areas, where residents more often displaced individual migrants than vice versa, but none of the researchers was able to demonstrate the exclusion of migrants from these habitats that is required for it to have meaning.

I spent two years studying wintering migrants during my thesis research in Texas and Mexico, as described above, and found no evidence of competition between wintering migrants and residents for food in

Veracruz tropical rain forest or Texas chaparral and riparian forest, contrary to the prediction of the Northern Home Theory that migrants in stable primary (unaltered) habitats would have to compete directly and overtly with resident ecological counterparts for food resources. In fact, I found that migrants clearly had their own niche space in their wintering environments, where nearly half of migrant species have resident (nonmigratory) populations that remain to breed during the summer months.

Once the "ecological counterpart" idea has been removed from the equation, and the understanding dawns that migrants live in—and in many cases have populations that breed in—their wintering environments, we can begin to understand the migrants' true nature and the essence of the migration phenomenon as a means for maximizing fitness by moving out of environments (the lower-latitude range from where they originated) where competition with others of the same species is intense for food, mates, and nesting sites into northern areas (current breeding range) where these resources are temporarily (during the summer months) in abundant supply.

The fact of the matter is that it is during the summer months in many northern areas when food is *not* limiting, which is why migrant species are able to move there from their wintering areas, why they show such broad breeding ranges across huge swaths of continents, and why they use such a variety of foods as the season progresses. It is during the winter months when they return to the places from whence many originated that food is most often limiting, which is why they require a specific niche in a particular habitat during this period, and why many show such restricted distributions. It is in these places where they evolved and where they can best compete for food and achieve their highest probability for survival.

What I found during my thesis work was that migrants of several species wintering in the neotropics arrive in suitable habitat (lowland rain forest in this case) in fall and establish a territory. Older birds return to the territory held the previous year while young birds must locate and establish one. If they are successful, they become "sedentary birds" and remain on this territory throughout the winter season. If they are not successful, they become "wanderers" and move into less suitable habitat nearby to feed and roost and continue to search in neighboring optimal habitat for an open territory, or they leave the area entirely to search for suitable habitat elsewhere. These wanderers suffer a much higher mortality rate than sedentary birds. Nevertheless, many will survive the winter season following this strategy, depending of course on the amount and suitability of the suboptimal habitat. For our Wood Thrushes in southern Veracruz, "suboptimal" habitat appeared to be young shrubby second growth, which apparently had rich patches of food but resources were less dependable and more dangerous to exploit. Of course, optimal habitat, such as lowland primary forest, was actually suboptimal for wanderers, as opposed to territory owners. In addition, higher elevation forest could be optimal in good weather and suboptimal during cold fronts (*nortes*).

Extensive work during the late '80s radio-tracking sedentary resident and wanderer Wood Thrushes on foot and by plane revealed that wanderers would spend part of their days in rain forest, evidently searching for undefended territories, and part in scrubby second growth foraging on undefended food sources. Wanderers suffered significantly higher mortality rates, mostly from avian predators, than sedentary birds.

My former graduate student and collaborator Kevin Winker found that wanderers had significantly higher fat reserves on average than sedentary birds among his wintering Wood Thrushes in Tuxtla Mountain rain forest. We found this to be true for several other species of territorial, wintering migrants during our Tuxtlas work. This finding seems counterintuitive to many who consider fat reserves as a measure of "condition," where heavy fat reserves supposedly equal "good condition" (meaning a higher probability of survival) and low fat reserves equal "poor condition" (lower probability of survival). This supposition is incorrect. Fat reserves are not a measure of "condition" in the sense defined. Rather, fat represents a method for storing energy when more food is taken in than is required to meet current metabolic needs. Its deposition is controlled hormonally through processes that are not well understood. What we do know, however, is that birds, and many other types of organisms, do not always store as much fat as they can whenever they can. On the contrary, they are genetically programmed to store fat at particular times or under specific environmental conditions. For instance, migratory birds spend a month or two during the postbreeding period eating only as much as is required to meet their daily activity needs, storing no extra fat. Then, a week or two prior to departure on migration, their behavior and physiology change: they eat intensively (hyperphagia), storing the excess energy consumed as subcutaneous fat, with individuals of some species adding as much as fifty percent to their body mass. Once stores are sufficient, they depart. After completing migration and arriving at their destination, their physiology and behavior change, and they no longer take in more food than their daily needs and no longer store large amounts of fat.

For territorial birds in winter, few individuals carry more than very light fat loads for several months until time for departure on northward migration nears. However, wanderers are confronted with a situation not dissimilar to that of a transient. In essence, they are in a state of perpetual migration until they are able to locate and defend a suitable territory, taking advantage of resource patches whenever they can find them and storing excess intake as fat. That, at any rate, is my hypothesis to explain why wanderers tend to carry more fat than territorial residents.

Another misconception exposed by Kevin's work is that density of individuals is an accurate measure of habitat value. In fact, as demonstrated by our Wood Thrush observations, density is often a measure of habitat instability, where large numbers of individuals are found at temporary food concentrations in contrast with low numbers of widely spaced territorial individuals where resources are stable for long periods. This concept is illustrated clearly in Stephen Fretwell's depiction of ideal despotic distribution for two habitats of differing quality: the lower-quality habitat can often have densities of individuals far higher than that of the higher-quality habitat—due to competition or lack thereof (discussed in detail in Chapter 8).

Jerram Brown presented a theory predicting that one would only see territorial behavior when a critical resource is in short supply and is economically defendable. This idea, discussed in detail in Chapter 8, helps to explain results from our field work in Mexico, along with results from later investigations of wintering-ground behavior in Mexico and elsewhere in the New World tropics regarding the various sociality patterns observed in migrants during the nonbreeding period.

1. *Solitary, Sedentary, and Territorial.* Both male and female members of at least twenty migrant species were found to remain on small individual feeding territories throughout the wintering period (October–April) in Tuxtla rain forest, and they returned to these territories in subsequent years. These territories were defended rigorously against intrusion by members of the same species—regardless of sex or age—by use of vocalizations, displays, and actual attack and fighting. Territorial behavior has been reported by a number of other researchers around the world in more than a hundred wintering migrant species, although most such studies lack data on within- and between-season persistence and site fidelity.

2. *Participation in Mixed-Species Flocks.* Not all migrants spending the winter in Tuxtla rain forest lived on territories. Members of several species joined mixed-species flocks. The mixed-species flock is a curious and much-studied phenomenon that occurs during the nonbreeding season in many habitat types and life zones, from boreal regions to the tropics, throughout the world (except, curiously, on some islands). One can find them in forests of the eastern United States during winter, where participants often include a few Black-capped Chickadees, a pair of White-breasted Nuthatches, a pair of Downy and Hairy woodpeckers, and a Brown Creeper or two, while in Amazonian rain forest there can be different flocks at various levels from ground to canopy, each composed of members of a hundred species or more.

The structure of these flocks is surprisingly stable in terms of the species composition for given periods of time (days, weeks, months), numbers and identities of individuals of each species, and home range. The chief advantage of participating in such a flock appears to be an increase in safety from attack by avian predators resulting from an increase in the number of observers keeping an eye out. Members of at least eight species of migrants, including such birds as the Black-and-white Warbler and American Redstart, participated in rain forest mixed-species flocks that included members of thirty or more resident species as well. The resident species participating in mixed-species flocks often occurred as male-female pairs or small family groups. Migrants, however, were usually the only individual of their species in the flock, and, in fact, would not allow other members of the same species to join the assemblage, essentially defending the flock space as a mobile territory, shifting in location as the flock moved through its home range. Our banding studies in the Tuxtlas showed that migrant members of mixed-species flocks remained with a particular flock throughout the winter season and returned in following years, although we did not have enough information to calculate return rates. We also found that Hooded Warblers would join mixed-species flocks that entered their territory and remain with them until the flock exited, a finding that Richard Chandler observed for Golden-winged Warblers in Costa Rica as well.

Based on our work done in the '90s in pine-oak habitat characteristic of the Central American highlands (above five thousand feet), migrants joined twenty or so species of tropical residents in mixed-species flocks. Golden-cheeked Warblers, a migrant that breeds in central Texas, usually occurred as single individuals, but in about ten percent of flocks there might be two birds, almost always a male and female. Other wintering migrants occurred in these flocks as well, but in numbers different from the Golden-cheeked Warbler: Townsend's Warblers were usually present in groups of less than four; Black-throated Green Warblers in groups of five to seven; and perhaps as many as ten Hermit Warblers were in these flocks. We had no data to confirm our suspicions, other than the small numbers themselves, but it appeared to us that flock participants for each species were not a random number; that is, that behavioral interaction likely limited the number of individuals of each species to what the flock space could support.

3. *Participation in Conspecific Flocks.* Individuals of many migrants occur in single-species flocks throughout the winter season. The work of Jerram Brown mentioned above predicts this type of behavior in cases where the distribution of the main foods required for survival occurs

in indefensible clumps. In such cases, association in flocks can have two benefits: first, the same kind of predator detection enhancement as occurs for members of mixed-species flocks; second, the collective experience of the individuals in the flock in locating and exploiting clumps. Examples of the kind of resources distributed in clumps include grain fields, schools of fish, invertebrates on a beach, fruiting trees, insect hatches, and so forth. Interestingly, such flocks often are composed almost entirely of members of a single sex for any given species.

4. *Solitary, Nonterritorial Wanderer.* This type of sociality is the most commonly noted resource-exploitation behavior for passerine migrants to Africa and the New World tropics. Indeed, as I have noted in previous chapters, observations of wandering winter migrants serve as the principal form of data from the nonbreeding period in support of the Northern Home Theory. Careful investigation of the other types of winter sociality provides some insight, however. Data from the Tuxtlas on Wood Thrush wintering behavior demonstrated that birds unable to obtain territories in forest tended to collect in scrubby second growth, where they moved apparently at random in search of temporary resource concentrations, suffering mortality rates significantly higher than those of territorial individuals in preferred forest habitat. Similarly, as in the case of migrants unable to find and defend territories in appropriate habitat, migrants presumably prevented from joining mixed-species flocks in forest were found as wanderers in scrubby second growth, likely suffering higher mortality in such circumstances than flock-following conspecifics in forest, although there are no actual data on comparative survivorship for members of the different resource/habitat use categories.

These findings suggest that the wandering strategy for wintering migrants is, in essence, a form of continued migration in search of more stable habitat in which to pass the winter season. This possibility appears clearer for whole cohorts of migrant species that appear to pass the early winter season as wanderers in the Sahel region of sub-Saharan Africa, where food is superabundant for a period of a month or two, before moving farther south into more stable habitats where they can pass the remainder of the winter.

In this context, the findings of Barbara Helm and Eberhard Gwinner with regard to onset and duration of frustrated movement (*Zugunruhe*) in stonechats (*Saxicola*) are relevant. They discovered that such activity, normally assumed to be associated solely as an indicator of migration, could be stimulated and maintained in caged migrants at any time during the annual cycle. This surprising fact has been found and reported by other researchers for other migrant species in which *Zugunruhe* could be stimulated by variations in external conditions, such as availability of food. Clearly, movement has advantages when food is absent.

The Importance of Sex during the Wintering Period

I have said that survival is the most important aspect of fitness for migrants during the winter period, a statement that provides the theoretical basis for much of the above discussion of sociality based on food resource needs. This obvious claim requires some qualification because what is sauce for the goose is not always sauce for the gander. The components of fitness (survival versus reproduction) are not the same for males and females, even if they are members of the same species. It is necessary to tread a bit carefully here because the entire subject of male dominance has a tendency to subvert logic. Nevertheless, let us proceed, albeit with caution. Let's start by looking at the data:

1. Mean timing of migration movement differs by sex for many migrants.
2. Mean distance of movement differs by sex for many migrants, with males generally wintering nearer to the breeding ground than females.
3. Proportion of males and females differs by habitat for a number of migrants.

One way to consider these data is through the lens of dominance, which is to say, males move at the optimal time to the optimal distance and choose the optimal habitat in which to winter, forcing females into less optimal situations through straightforward superiority in competition. As summarized in *The Avian Migrant*, many researchers interpret these data in this way.

There is, however, a problem with this conclusion, namely that if males dominate females in competition for critical overwintering food resources, then female survival should be significantly lower than male survival. Several authors have claimed that this is true based on return rates of males and females to breeding territories occupied the previous year. In general, male songbirds are twice as likely as females to return to the previous year's territory. However, there is a serious problem with this reasoning (as demonstrated by careful studies of migrant breeding populations, such as that conducted by Val Nolan in his study of the Prairie Warbler), namely the assumption that males and females return at the same rate to the previous year's territory. Nolan found this assumption to be wrong. While male Prairie Warblers nearly always return to occupy their breeding territory of the previous year, many females, although they do return to the same region, do *not* return to pair with the same male (as discussed in Chapter 6). For the purposes of this discussion, it is sufficient to emphasize that there are no good data to support the idea that male annual survivorship exceeds female annual survivorship for any migrant species of which I am aware.

That being the case, I suggest we need a different explanation for the differences observed in male and female wintering ecology and behaviors such as protandry (early male return to breeding area). One such difference is the obvious one suggested by high male return rates, namely that territory ownership is key to male reproductive success, and the best chance of establishing ownership is return to occupy the territory of the previous year. The situation is different for females. By returning a week or so later on average than males, they can choose among the territorial males for the one with the best territory, which may or may not be their mate from the previous year.

Looked at in this way, one can see that what is optimal in a fitness sense for the male, namely a wintering area that is as near as possible to, or even on, the breeding territory for the likelihood of increased reproductive success even at the cost of lower survivorship, might not be optimal for the female, who has no need to factor reproductive success into the equation when considering optimal wintering habitat.

But what about migrants like the Hooded Warbler in the Tuxtlas? Territorial Hooded Warbler males outnumbered females in rain forest eight to one, while females were more often found on territories in thickets, where they outnumbered males. Females appeared to be excluded for the most part from holding territories in more open understory of the forest, but they appeared to be perfectly capable of defending thicket territories from males. But was female winter survivorship lower than that of males throughout the range? Certainly there was no evidence of that in terms of numbers of birds on northbound migration in our netting samples in South Texas (where the bird neither breeds nor winters). Females outnumbered males in those samples in the spring of both 1974 (17:12) and 1975 (78:44).

There are at least four ways in which females could maximize nonbreeding-season survival in competition for food with males (which are often larger or more brightly colored):

1. *Be exactly like males in terms of size and appearance.* The vast majority of tropical species have sexes alike in appearance, at least to a human observer (I'm pretty sure no male White-breasted Wood-Wren lurking through the jungle understory would mistake a female sashaying around her bower for another male). While most males and females are similar in resident birds, males are more brightly colored than females on average in about a third of migrant species. As discussed in Chapter 6, "Breeding Period," the likely reason for male ornament in migrants is to improve chances of success in competition with other males for high-quality breeding territories. Indeed, males of some migrant species, like the Scarlet Tanager, molt out of their bright plumage in fall, after the need for a bright plumage has passed, into less conspicuous attire, molting back in spring prior to northward migration. Nevertheless, males of many migrant species retain their brighter coloration throughout the winter, and in members of these species, like the Hooded Warbler, there is no question that the brighter

FIG. 4.6. Hooded Warbler plumage variation: "normal" female (*left*); "different" female (*center*); "normal" male (*right*). PHOTOS (*LEFT TO RIGHT*) BY MARY HALLIGAN, LOTUS WINNIE LEE, AND BARTH SCHORRE

plumage provides an edge in terms of the effectiveness of displays used in defense of winter feeding territory as compared with females with more cryptic plumage (see Figure 4.4, above).

So, what's a girl to do? One option (in an evolutionary sense, of course) is to look exactly like a male. The fact that most migrant species with brightly colored males are called "dimorphic" (bright male plumage, cryptic female plumage) is evidence that this option must have some costs in terms, perhaps, of higher predation rates. However, specimen data demonstrate that the situation is not quite so simple. My own work with several supposedly "dimorphic" species has revealed that they are not dimorphic—they are polymorphic, in the sense that there are not two plumage types, a male and a female, but several plumage types: a "normal" brightly colored male and a "normal" cryptically colored female, along with percentages that vary from species to species of females that look similar to males (andromimesis) and males that look more like females (gynomimesis) (Figure 4.6).

I have found evidence of this type of polymorphism in twenty other supposedly "dimorphic" species, including the Baltimore Oriole and Canada Warbler, and would not be surprised to discover its occurrence in all such species. I would also not be surprised to discover that percentages of male-like females are changing as competition for winter habitat becomes increasingly intense, with destruction of tropical forest at over ninety percent for most parts of Middle America.

2. *Winter at different latitudes.* Colgate University is a small liberal arts college in central New York (it had 1,600 students, all male, when I went there). It is basically a teaching institution, but a research endocrinologist, Dr. Roger Hoffman, joined the biology faculty during my junior year. I enrolled in his "problems" course and pursued a study of the effects of drugs on perversion of American Tree Sparrow behavior—testosterone for females and estrogen for males. However, I quickly found that, although the species is a common winter resident in the region, there were no males among the birds that I was able to capture. I thought that I had made a remarkable discovery, but a cursory review of the literature revealed that the phenomenon of differential wintering by latitude for the different sexes was common, not only in American Tree Sparrows but also in nearly all species that migrate less than a thousand miles and even in some long-distance migrants. Adult males of these species tend to remain on or near the breeding territory, or they move much shorter distances away than females or juvenile males—a well-known fact for more than a century. Whether males choose the optimal migration distance and force females and juveniles to move farther, or whether it is the females that choose the optimal migration distance while males accept greater risk of decreased winter survival for increased probability of maintaining a higher-quality breeding territory, is unknown.

3. *Use of different habitats or different parts of the same habitat.* I have mentioned that wintering Hooded Warbler males outnumbered females by eight to one in open rain forest, whereas females outnumbered males in thickets. My former Smithsonian colleague Pete Marra found a similar situation in American Redstarts wintering in Jamaica, where males outnumbered females in mangroves and females outnumbered males in scrub.

Eastern Phoebes wintering at our study site at Welder also show differences in habitat use by sex. Phoebes, which breed across much of the eastern United States, occur only as winter residents at Welder, where they begin arriving in mid-October, establishing small territories in woodlands and thorn forest bordering creeks and roads. I was able to map out territory boundaries using playback for those found along the small tributary running from behind our house a few hundred yards down into Moody Creek. There were seven such territories. Once the boundaries had been mapped, I went out each day and removed the owner. Often, by the time I went out again the following morning, a new owner had taken over, which I also removed. I continued this regimen for about a month until shortly before our time to leave for Mexico. Ultimately, Territory 1 had three different owners; Territory 2, eight; Territory 3, four; Territory 4, six; Territory 5, three; Territory 6, five; and Territory 7, two. Clearly, winter territories were important, and early in the season, there were more phoebes than suitable territories. Additionally, I tried to take territory owners that were singing as well as chipping to find out if females were using song in defense of territory. Eventually ten singing birds were taken, five of which turned out to be females.

The phoebe territories occurred in two different habitat types: thorn forest and riparian forest borders. Use of these different habitats for territory location by sex for phoebes was not random. I removed fourteen male and four female phoebes off territories in mesquite thorn forest and eleven female and nine male birds off from territories in riparian forest. Statistically, these proportions are significantly different.

In fact, in all of the species where I have been able to compare male/female ratios in specific wintering habitats based on specimens, which isn't that many, I have found significant differences in distribution based on sex.

Dr. Marra, and several other investigators, explain these differences as a result of competition in which males choose optimal wintering habitat (higher probability of survival) forcing females into suboptimal habitat (lower probability of survival). For reasons explained in detail in *The Avian Migrant*, I believe that differences in wintering habitat use between the sexes may result from differences in what is "optimal" for males versus females. The only way to test which of these explanations is correct is to compare male versus female survivorship for birds wintering in the different environments. I am not aware of any migrant species for which this has been done.

4. *Travel in separate flocks (for those using clumped resources).* In late December of 1982, a classic "blue norther" hit South Texas, dropping tempera-

tures below the freezing mark for a week. This disaster killed many tropical plants in the region, including the large bougainvillea growing on our balcony in Kingsville, most of the *Washingtonia* palms lining boulevards in Corpus Christi and Brownsville, and nearly all of the seventy thousand acres of orange and grapefruit trees growing in the Rio Grande Valley.

About half of the citrus acreage was replanted, and by early 1986 many of these young trees were producing fruit. It was then that a curious thing happened. Great-tailed Grackles, which previously had only one annoying habit, namely that of congregating in roosts in town squares, pooping on strollers and screeching all night, developed a second. They invaded citrus groves, pecking the developing fruit.

You can be forgiven for not understanding why this pecking was a problem, but it was huge for the Valley citrus industry, still reeling from the devastating winter of '83. The issue is that peck marks cause an orange or grapefruit to be reduced from "table" grade (that is sold for individual consumption) to "juice" grade, losing half its value or more. The Valley Citrus Association, a consortium of growers in the region, estimated that the damage from this pecking was in the millions of dollars, and they turned to the federal government to *do something about it*. Fortunately for them, the chairman of the House Agricultural Committee at the time was their local congressman, Kiki De La Garza, and Kiki was quick to add a million-dollar appropriation to study the issue to Federal Animal Damage Control's budget.

I was a research scientist at the Caesar Kleberg Wildlife Research Institute at Texas A&I University (now Texas A&M University) in Kingsville when this happened, and our director, the late Sam Beasom, used his considerable political skills to land us the contract for the study. He assigned me and my buddy Alan Tipton to be lead investigators; we were quickly dubbed Pancho and Lefty after the characters in the Townes Van Zandt song of the time for reasons that are obscure (maybe because, although Alan was not left-handed, he was quirky—a thoughtful mathematician trained in statistics and modeling with a PhD from Michigan State—and I was superficial, more apt to "shoot first" than think when confronted with field difficulties).

We assembled an excellent team of a dozen or so graduate students and technicians (including my son John Jr. for one summer), and we launched a three-year investigation of all things grackle. We set up a base camp in a rented trailer in the city of McAllen for team members to use when experiments were underway in the groves, and Alan and I made the two-hour commute from Kingsville often, usually meeting the boys at the 10th Street Whataburger (five cent coffee with your Whataburger cup) for breakfast to discuss the latest experimental results and devise new ways to save the Valley citrus industry from the grackle menace.

We had a great time, won a major award—President Reagan gave us a Golden Fleece Award in 1988, as reported in *Time* magazine, for wasting taxpayer money (Kiki was a Democrat)—and found out a lot about our adversary.

The main discovery of this work was that it takes a lot to keep grackles out of groves. None of the various methods advertised by "wildlife protection for crops" companies worked—not timed canons, not monster balloons, not mothra kites, not plastic owls, not predator calls, not silvery streamers. The only thing that worked to get a flock out of a grove was to start shooting at them with a shotgun, and that only served to move them to a neighboring grove, and only during the winter. We found no way to get them out of a grove once nesting had begun.

Our most exciting finding from the perspective of our erstwhile clients, the Valley Citrus Association, was location of a winter roost in a thirty-acre sugar cane field bordering the Rio Grande just outside the town of Mercedes. We discovered this roost by capturing a number of grackles during the day, equipping them with radio transmitters, and then following them home to bed in the evening. What was so thrilling for our clients was that the roost held nearly all of the grackles in the middle Rio Grande Valley— an estimated 1.5 million birds. This discovery set off a series of frantic calls between our local Animal Damage Control collaborators and their superiors in Washington to see if a "nuclear option" might be possible—that is, if a plane could be dispatched loaded with a chemical that could wipe out these expensive thorns in the citrus industry's side in a single night. For-

FIG. 4.7. Female (*left*) and male (*right*) Great-tailed Grackles. PHOTO BY R. K. STEWART

tunately, it was decided that the potential benefits of a grackle-less future were outweighed by the potential costs for the good people of Mercedes, who, by the way, were not consulted regarding this momentous decision.

The relevance of this digression for our present discussion of sexual differences in winter food use is that each morning, when our marked birds set out from the roost to head for the various groves and agricultural fields in the vicinity (anywhere within a thirty-mile radius of the roost), they traveled in mostly single-sex flocks of a few to several hundred birds, which traveled to different feeding destinations. We also found that these flocks were not cohesive from one day to the next. Each individual formed its own decision daily as to which flock to join; maybe joining a flock on its way to a melon field five miles to the north one morning and one headed to an okra field eight miles to the east the next. The only constant was that females generally joined flocks of females while males joined flocks of males.

Why?

A Dominance Theory adherent might suggest that females don't want to associate with males because males are twice their size and likely to win any direct confrontation over any particular morsel (Figure 4.7).

This explanation makes no sense for me. If male dominance at feeding sites was the explanation for "female only" flocks, what's to prevent males from joining female flocks? Also, thinking of Jerram Brown's cost-benefit analysis for territorial behavior (Chapter 8), bullying females for bugs in an open field seems like a waste of time and energy. A male may prevent a female from eating a specific bug, but so what? She just moves a little ways away and eats another one just like it.

I like the idea put forward by Peter Ward and Amotz Zahavi in a paper published in 1973, wherein they suggest that flocks departing from roosts can serve as *information centers*. They suggest that individuals join flocks as they form up and leave the roost because they think flock members know where a good food source is located. Following this reasoning, males join males and females join females because what constitutes "good" for a male may not be the same for a female, especially if they differ significantly in structure.

Conclusions

I think that key factors to consider in comparing male versus female winter ecology and distribution is that the sexes in all migrants of which I am aware differ at least somewhat structurally (in size and often in plumage coloration) and in terms of behavior. These differences likely require different balances to be struck between reproduction and survival, and until we know more about what these balances are, and their various consequences in terms of fitness, it is probably useless to speculate about differences in dominance or competitive ability during the nonbreeding period. One thing we can be quite certain of is that if being different from a male did not have fitness benefits as well as costs, then natural selection likely would assure that females were exactly like males in all features affecting competitive ability for food, as is true for over seventy percent of migrants. There is no reason to believe that females are condemned to a subordinate status simply because of their sex.

CHAPTER 5

SPRING
Migration

Spring migration is where it all begins. As stated repeatedly, migration starts with dispersal, and the purpose of dispersal is to locate a place where the young bird can reproduce. Birds leave their wintering areas, which for many are the same areas in which they evolved, in order to raise more offspring than would be the case if they remained to breed where they spent the winter.

Departure

> We sailed on the good ship Venus,
> My God you should've seen us!
> —"Good Ship Venus" (Traditional)

It is interesting to note that while weather is almost always cited as the cause of fall movement of birds, it is seldom mentioned with regard to initiation of northbound migration. The reason is obvious. What changes in weather that do occur for the wintering areas of most species in spring actually signal *increase* in foods available for many, which is why populations for nearly half of New World temperate zone migrants remain in those areas to breed. Nevertheless, over half of them do leave *despite* a complete lack of declining food resources. I have proposed that the reason for this behavior is that those that leave can produce more offspring than those that stay.

In any event, after spending several months on their wintering grounds, migrants mysteriously begin to prepare to move back north. Even less is known about preparation for the spring migration than about fall migration, and nearly all of that knowledge comes from laboratory experiments with caged birds of a relatively small number of species. These experiments demonstrate that initiation of preparation, as indicated by intensive eating and deposition of fat reserves, is under the control of an internal clock that appears to be set based on some aspect of photoperiod during the preceding summer. This finding explains how birds are able to time their return to temperate breeding areas properly even when wintering in areas on the equator, where no change in photoperiod could signal proper timing. Because these birds appear to be genetically programmed to begin preparations for return to the breeding ground on a particular date, they often are referred to as "calendar" migrants, in contrast to species that seem to begin preparations more in response to local conditions, which are referred to as "weather" or "facultative" migrants.

In truth, however, the differences in timing control between these groups are not quite so clear-cut. Some of the species on which laboratory experiments measuring timing of commencement of preparation for spring migration were performed were, in fact, weather migrants, which would appear to demonstrate that a genetic "calendar" component was part of their migration timing, perhaps in addition to an environmental component.

Regardless of whether they are calendar or weather migrants, males leave the wintering ground on spring migration a week or two earlier than females for most songbirds. This behavior is known as "protandry" in the literature, and it is often cited as additional evidence of male dominance over females in terms of access to critical foods. The argument goes like this: Through behavioral dominance, males of several species (such as American Redstarts) establish territories in the "best" wintering habitats (in terms of quantity and quality of food supply), forcing females to winter in more marginal habitats where food sources are poorer or less stable. This behavior results in superior ability to obtain the resources necessary to prepare for departure, which allows males to depart at the optimal time for northward migration. Female departure is delayed to a less optimal time by their inability to compete with males for critical resources, requiring them to wait until males have left to obtain them.

As discussed in the chapter on migrant breeding biology (Chapter 6), there are other reasons why females might delay arrival until after males. Be that as it may, a group of researchers led by Jessica Deakin recently tested the "male dominance" hypothesis explanation for protandry in Black-throated Blue Warblers. They captured males and females in fall and held them in captivity over winter. Although provided the same diet and photoperiod stimulus, males initiated spring migratory restlessness about a week before females, demonstrating that, at least for this migrant, dominance had nothing to do with wintering-ground departure timing. Similar experiments have been performed on stonechats by Barbara Helm and her associates with similar results.

Route

> Alice: Would you tell me please, which way I ought to go
>   from here?
> The Cheshire Cat: That depends a good deal on where
>   you want to get to.
>   —Lewis Carroll, *Alice's Adventures in Wonderland*

The term "route" is really a misnomer. It implies some sort of highway, like Route 66. Migration routes have no such physical reality. They are, in fact, nothing more than mathematical frameworks based on statistical probabilities, that is, a span of land where the largest number of individuals of a given species have been sighted during migration relative to other areas. OK, you may say, so what? It helps people to visualize the bird's travel. That's true. But it also implies that it exists as some sort of construct in the bird's mind, and here I beg to differ. I think no such construct exists, and that the data we have does not support its existence. What I believe exists in the bird's mind is a sense of destination, perhaps modified by an inherited sense of the best directions to take to get to the destination (breeding ground or wintering area) in a given season. Somehow—I do not understand how—the bird knows where it's supposed to go. In addition, it has all of the navigational skills necessary to get there. I also believe that this is a characteristic common to many forms of life, those that are resident as well as those that are migrant. Thus, when that first resident bird sets out from where it was hatched to try to find a new place, it will always know how to get back to where it started from, and for the offspring of migrants, they too will somehow know how to get back to where their parents started from, even though they themselves have never been there.

German scientists at the Radolfzell lab built a complete theory explaining how a novice juvenile bird on its temperate breeding ground can navigate correctly to its appropriate wintering area, despite never having been there, based entirely on a genetic program for route. Their theory posits that the bird is born with a genetically inherited sense of the correct direction to take to get from the breeding area to the wintering area and, in addition, an inherited sense of the amount of time it should proceed in this direction in order to arrive at the wintering area. This hypothesis is called the Migration Route hypothesis, and it is based on those classic "time and distance" problems from seventh grade math (based on personal experience in Miss Dunn's class at Washington Junior High School in 1959). "If you leave Utica at eight in the morning, traveling by barge at a rate of three miles an hour, at what time will you arrive in Buffalo, two hundred miles to the west?" Simple calculation. A child can do it, and a bird can do it too—if it has a program for the right direction to head in, and for how long it's supposed to travel. With these genetic gifts, birds can find the proper wintering area without ever having been there.

Whether or not this hypothesis is correct, or, if true, to what extent modifications imposed by environmental or personal circumstances can occur, remains conjectural. Regardless of the degree to which route is un-

derstood to be part of programmed behavior, when we began our investigations of migrants in the early '70s, the paradigm was that most species
followed the same route north as they followed south. As it happened, our
work ended up providing some insight into this idea.

Over the course of our two years of mist-netting in Texas and Mexico,
my late colleague, Mario Ramos, and I, along with the assistance of several outstanding helpers, gathered an immense amount of information on
volume of fall and spring migration for passage migrants at our two study
sites on the west coast of the Gulf of Mexico, located almost eight hundred
miles apart. A striking result from our Texas netting of transients was huge
differences in fall and spring migration volumes for some species (Table
5.1). Even more remarkable was that volumes for a number of these were
similarly skewed in Mario's samples for southern Veracruz. Neither Mario
nor I had included volume of migration in our initial thoughts on what
we might learn, but in summarizing capture data for our theses, it became
obvious to both of us that something interesting was going on.

We had the opportunity to take a closer look at these data when we were
invited to prepare a paper for the First Welder Wildlife Foundation Symposium, scheduled for October 14, 1978, in Corpus Christi. We justified (to Bonnie and to Mario's wife, Isa) the time in preparation and expense (we would
have to travel from Minnesota to Corpus to give the paper at our own expense) as worth it professionally because the proceedings would be published.

Accordingly, we pulled out all our catalogs and gathered what literature
we could find on fall and spring migration volumes from elsewhere around
the Gulf, from Florida to Texas, and met several days and evenings over the
dining room table at my house in St. Paul for three weeks or so, and we put
the thing together.

As we analyzed the information, we came to realize a couple of profound truths regarding songbird migration in North America, namely that
each species had its own migration pathway shaped mainly by relative location of the breeding and wintering areas, and that the route followed to
the wintering ground in fall was quite different for many species from that
followed on return to the breeding area in spring.

TABLE 5.1.

Total migrant captures by season for our work at Welder on the Texas Gulf coast

SPECIES	FALL 60,000 net-hours	SPRING 35,100 net-hours
Acadian Flycatcher	0	188
American Redstart	21	71
Bay-breasted Warbler	2	62
Black-and-white Warbler	0	207
Blackburnian Warbler	2	32
Black-throated Green Warbler	7	57
Blue-winged Warbler	5	56
Canada Warbler	245	476
Chestnut-sided Warbler	2	122
Common Yellowthroat	12	232
Golden-crowned Kinglet	0	25
Golden-winged Warbler	0	27
Gray Catbird	26	110
Gray-cheeked Thrush (includes Bicknel's)	1	74
Indigo Bunting	23	64
Kentucky Warbler	19	240
Louisiana Waterthrush	6	24
Magnolia Warbler	13	279
Northern Waterthrush	41	204
Ovenbird	38	505
Rose-breasted Grosbeak	1	20
Summer Tanager	1	27
Swainson's Warbler	4	48
Tennessee Warbler	16	260
Veery	4	31
Wood Thrush	11	79
Worm-eating Warbler	2	118

We considered these data to be of the first importance. We both were familiar with the concept of differences in major migration pathways from the classic 1915 publication by W. W. Cooke, and we both knew that Cooke had mentioned that at least one species, the American Golden-Plover, was known to follow a different path south to its southern South American wintering grounds than it followed north to its Arctic breeding grounds. What we had not recognized was the diversity of routes and the likelihood that most individual migrants followed their own routes, not some pre-programmed genetic map. In addition, most appeared to follow a different way north from what they followed south for the simple reason that environmental conditions (wind, weather, food resources, etc.) differed between fall and spring, thus favoring different routes.

We gave the paper at the small Welder conference, but we weren't happy. True, it would be a publication, but who would read it? We were both busy with other stuff, but we kept our eyes open for another opportunity to get the information out to the broader ornithological community, and we thought we had one when our paper was accepted for presentation at the American Ornithologists' Union meeting in College Station, Texas, August 13–16, 1979.

Accordingly, Bonnie and I attended and I gave the talk. Afterward, the assistant editor for the journal *American Birds* came up, congratulated me on the stimulating information presented, and asked me to submit a paper for publication in the journal. I was very excited. Invited submissions are rare (I've received only four or five since—usually to allow response to vituperative comments from compeers), and although it was not a top-of-the-line professional outlet, it had a broad ornithological readership, and our ideas would be out there in the intellectual marketplace.

I was overwhelmed with other matters for the next several months, but the opportunity finally came to devote the time needed to the paper when Bonnie became ill with preeclampsia in June 1980, about two months before our third child, Nathaniel, was due to be born. She went into the hospital then in hopes of carrying him closer to ideal delivery date without damage to either of them. I spent most days with her over the next three weeks until Nathaniel was delivered on the July 9, five weeks early, by emergency C-section. Although stressful for all, both Bon and Nate came through OK, and I had had the time to do the work for the *American Birds* paper.

In performing the literature review, I discovered an interchange on the topic of migration routes across the Gulf of Mexico. The January 1945 issue of the ornithological journal *The Auk* contained an article by Rice University English professor George Williams, in which he presented data on spring migration volumes along the Texas Gulf coast for several species of migrants. These data appeared to document a different route north for these species, passing along the western edge of the Gulf of Mexico ("circum-Gulf") from that followed south across the Gulf of Mexico ("trans-Gulf") in fall. This finding challenged the existing paradigm for North American migration routes as established by W. W. Cooke in his cardinal 1915 disquisition on the topic. Therein, Cooke described and pictured five major migration routes for the continent, including a trans-Gulf route. For all of these, Cooke assumed that migrants followed the same path north as they followed south, with the possible exception of the North Atlantic route (known to be followed by the American Golden-Plover and some other shorebirds), which seemed to involve a more westerly, continental route northward in spring.

Williams knew that in presenting his theory of seasonal change in migration routes he was challenging an icon, George Lowery, professor of zoology at Louisiana State University and leading migration route theorist at the time. Therefore, Williams was at some pains to make his case watertight. This effort resulted in unfortunate overstatement. What he knew was that many species of migrants occurred in numbers many times greater in spring along the Texas coast than in fall. This piece of knowledge was not available to Cooke when he drew up his maps because few observers had ever reported on migration along the Texas coast at the beginning of the twentieth century. Had Williams restricted his paper to reporting on this phenomenon, he might have been OK. But in trying to cement his case, he cherry-picked the literature to try to prove that *all* migrants that were trans-Gulf in fall were circum-Gulf in spring, which is false. This approach brought him directly in conflict with Dr. Lowery, who, from his lofty perch at LSU, had made a detailed study of both fall and spring, trans-Gulf migration. Heated pronouncements and rebuttals between Williams and the LSU crowd ensued in the literature over the next several years, but the issue was never in doubt. Lowery and his boys eventually squashed the circum-Gulf theory like a bug.

I found this intense debate fascinating, especially since I now knew that Mario and I had not discovered seasonal route change for trans-Gulf migrants. That honor belonged to George Williams. Nevertheless, we knew that while Williams was wrong about a lack of spring migration over the Gulf, he was right about spring routes being different from fall routes for a majority of Gulf migrants.

The paper resulting from my hospital interlude included a summary of the controversy in the literature as a lead-in to presentation of our data from Texas and Mexico, which showed huge differences in volume of fall and spring migration for many songbird migrants, with vastly higher numbers in spring than in fall, as shown for the Texas work in Table 5.1, above. We believed our findings to be perfectly straightforward. All we were doing, after all, was reporting actual data from two years of intensive fall and spring sampling along the western Gulf coast on a scale never before (or since) accomplished. What could be more innocent? We were therefore surprised, stunned actually, when Robert Arbib, editor of the journal *American Birds* at the time, sent us a letter rejecting the paper his editorial staff had invited. Included with the rejection letter was a five-page rebuttal of our route-change hypothesis by an anonymous reviewer. Even though we had made no claim concerning absence of trans-Gulf spring migration, documenting the obvious nature of a western swing of routes was obviously verboten as representing a challenge to the sanctity Dr. Lowery's trans-Gulf hypothesis.

We sent the paper to two other journals in attempts to get it published, only to have it blocked by a reviewer each time. Finally, fourteen years later, we were able to get it into print in a European journal, *Bird Conservation International*. Nevertheless, the idea that many individuals of many migrant species follow a different route in fall as opposed to spring has not found its way into standard migration thinking.

I was reminded of this a few years ago when I attended a lecture by a leading investigator of migration routes using the new, at the time, technology of data loggers. These tiny devices can be attached to a bird captured on the breeding ground in summer and when retrieved from the same bird the next summer provide information of a crude sort (give or take a hundred miles or so) on the route followed south to the wintering site in fall and the route back to the breeding site. The researcher showed data on four Wood Thrushes tagged in this way in northwestern Pennsylvania. Two of the four birds showed a route across the Gulf in fall and a westward swing of several hundred miles in spring. When I asked her about this, she said she thought it was simply an anomaly. In fact, she expressed surprise when I pointed out that there was a body of literature discussing just this phenomenon.

Several papers in addition to ours have been published on differential fall and spring routes for migrants from western North America, the Mediterranean region, Africa, and central Asia, yet none was cited in Ian Newton's exhaustive summary of the migration field, *The Migration Ecology of Birds*. His only comment on the subject that I could find was that "in some species, patterns of fat deposition, flight lengths and speeds of travel differ in a consistent manner between the two seasons, associated with different travel routes and stopping sites," a statement unsupported by any literature citations whatsoever.

The final word on this subject has been provided by the extraordinary sightings maps available for anyone to see at the eBird website (https://ebird.org/science/status-and-trends). My friend Jim Berry, retired executive director and president of the Roger Tory Peterson Institute, recently brought them to my attention (December 4, 2018). These maps show the total number of sightings each day throughout the annual cycle for about 120 species so far, among which are several that Mario and I found,

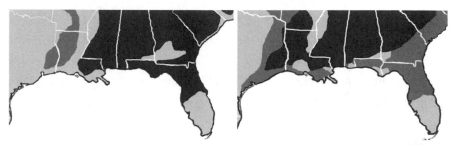

FIG. 5.1. Wood Thrush spring migration density (*left*) versus fall migration density (*right*), based on sightings, demonstrating a westward shift in route for millions of birds in spring. Black = high density; dark gray = moderate density. DATA ARE FROM EBIRD (HTTPS://EBIRD.ORG/)

based on our netting in Veracruz and Texas, to follow dramatically different routes in fall versus spring. In the Wood Thrush, for instance, sighting densities show a westward shift of hundreds of miles when spring migration is compared with fall (Figure 5.1).

As may be obvious to the reader, I am a little sensitive regarding this topic. However, there are important reasons for venting my spleen. My problem, I guess, is that my experience with migrants argues that migration route is extraordinarily flexible, not simply a genetic program. This flexibility, in my view, results from the fact that resources along vast areas of the potential routes are superabundant. A second factor is that migrants have incredible flight capabilities. The Bar-tailed Godwit, a shorebird, has been documented by data logger to fly eleven thousand miles nonstop. Even tiny Blackpoll Warblers can tank-up and fly seventy hours from the coast of New England to South America. Taken together, these facts mean that factors in addition to food normally shape the route, such as wind and weather, and these can be drastically different in fall from spring.

This finding of a westward shift in spring route, as compared with that taken in the fall, in perhaps a hundred or more species of trans-Gulf migrants clearly implies some inherited sense of the need (in a fitness sense) to modify the northbound route relative to that taken south. Seasonal differences in winds, storm probabilities, and food availability can make sharp differences in chance of survival between different routes. Thus, there is evidence of an inherited influence on route, as maintained by the Germans, but with a significant ability for the individual to respond to environmental circumstances to modify its route, enabling successful completion of its journey to ancestral breeding or wintering area.

Stopover

During field work at Welder during the spring of 1975, we had a golden opportunity to investigate details of stopover behavior and physiology for transient Northern Waterthrushes. Arriving back from our Mexican field work on March 18, I got net lanes cleared, and I had nets back up by the end of the month with help from Chris Barkan, who had volunteered to assist me through the spring season. Reconnaissance revealed a remarkable change. Although the mott was mainly a forest of hackberry, cedar elm, anacua (a tropical mid-story tree species), and pecan, an open, marshy swale now extended down the middle of it, perhaps two hundred feet west of the Aransas River banks. This swale had been mostly dry when we began work in the fall of 1973, but after deluges that occurred in October 1973 and September 1974, it formed a large, temporary pond bordered by extensive mud flats, twenty to thirty feet in width. These mudflats had not existed during previous field seasons, but they now provided perfect foraging habitat for transient Northern Waterthrushes. When I realized this, I decided to turn over all of the netting duties in the mott to Bonnie, Chris, and Karilyn Mock while I would focus on the extraordinary opportunity presented by the mott's temporary pond (Figure 5.2).

From April 10 until the May 16, I had ten nets with one net every fifty feet or so set around the periphery of the pond, each extending from the water's edge across the mud toward the forest. These nets were run only to catch newly arrived waterthrushes or recapture specific individuals at specific times. A total of 122 waterthrushes were captured, weighed, and color-banded during this period. Each day during the month or so of intensive field work at the pond, I captured or recaptured waterthrushes frequenting the pond edge, recorded weight and fat content, and then released them.

FIG. 5.2. A portion of the temporary pond in the middle of Hackberry Mott showing mist nets in place to capture Northern Waterthrushes on their mudflat territories, April 1975.

A deer blind atop a fifteen-foot stand was put up on the east shore; from this blind, it was possible to survey the entire pond and its muddy banks. I spent a total of sixty-two hours in that blind over the five-week period, recording interactions between individual waterthrushes, including displays, vocalizations, and chases, and I constructed daily territory maps for each individual.

As I have mentioned above, a territory is a fixed area from which intruders are excluded by some combination of advertisement (such as scent or song), threat, and attack. The existence of a fixed area actively defended and advertised is strong evidence that the social system is based on territoriality. Transient Northern Waterthrushes stopping to rebuild fat reserves at Welder exhibited this type of system. Males and females held separate territories, which each individual defended against intrusion by either sex. The primary means of defense was advertisement, for which the "chink" call note was used an average of eighty-six seconds per hour. Territorial neighbors did not normally cross each other's boundaries, so that most hostile activity was directed toward nonterritorial intruders. Attacks on these birds were frequent and usually turned into chases when the intruder fled. If an intruder did not leave when it was attacked, a fight ensued which involved physical contact. When two birds were foraging on the ground and happened to confront one another, a visual display was often given by one or both birds. Territories were held an average of 3.2 days.

Waterthrushes captured anywhere in the mott were released at the pond. Twenty of the 122 individuals captured were recaptured at the pond one day or more after release. Average weight change for these birds was +0.4 grams. For the six birds that were recaptured away from the pond area, the average weight change was +0.3 grams. Seven eventual territory holders were "floaters" (nonterritorial intruders) for at least a day prior to attaining a territory, and some individuals left the pond for a day or two before returning to establish or reestablish a territory. Two birds were forced out of territories that they had earlier defended. Territories changed in size and shape according to the number of individuals and their relative competitive ability. Bird #113 at one time held the whole pond, though it used only a part of it. This behavior was probably due to the relatively poor competitive ability of #112. When the pond was crowded, nonterritorial birds usually attempted to claim territories by defending an area at the border between two territories. They then expanded outward from this area; in two cases, they succeeded in forcing out the original territory owners.

Interlopers found territory boundaries by trial and error, entering a territory repeatedly and being chased out. They found that they were chased less often or not at all at the territory boundaries, and at these weak points, individuals were twice observed to set up their own territories.

Known territory holders showed a mean gain of 0.68 grams (about five percent of body weight) over the course of their stay, while other recaptures showed mean weight loss of 0.25 grams. Some birds that had territories left before achieving heavy fat loads, while others stayed at virtually the same weight after achieving heavy fat before leaving. Weights were taken as close to twenty-four hours apart as possible. Maximum daily weight gains were on the order of about seven percent of body weight. Some birds arrived with

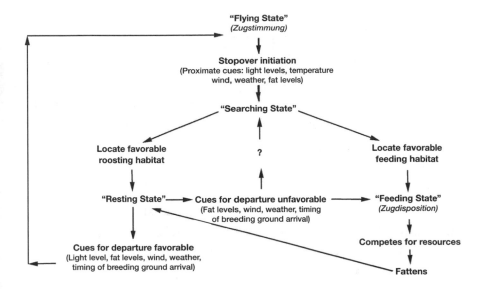

FIG. 5.3. Sequence of physiological and behavioral transitions between actual migratory flight and stopover followed by transients as they move from wintering to breeding area in spring, or from breeding to wintering area in fall. The question mark (?) highlights the hypothetical status of the "Searching State" (also called "Transit State").

fairly heavy fat, presumably more than enough to continue their journey, but stayed for several days. Thirty-eight individuals arrived with light or very light visible fat loads and disappeared without being captured or seen again.

As these experiments show, individual transient Northern Waterthrushes used a territorial system to guarantee critical resources for fattening. Average length of stay for territorial birds was 5.6 days but average time to hold a territory was only 3.2 days. I interpret this to mean that in most cases at least two days passed before a bird could gain a territory. Whether or not a bird was in possession of a territory had profound effects on the bird's weight: in the days prior to obtaining a territory, birds showed weight losses, while they showed significant daily weight gains after obtaining a territory.

Based on data collected during stopover for the Northern Waterthrush at Welder, plus outstanding additional research on transient birds collected by Swiss and German scientists, it is possible to construct a hypothetical scenario regarding the sequence of alterations between the various physiological and behavioral states involved in migratory movement between breeding and wintering areas, as shown in Figure 5.3.

In plain language, the migrant starts the sequence, either on the breeding or wintering area with the "Feeding State" (right-hand side of the figure). Once reserves are sufficient, it stops foraging and goes into a "Resting State," perched quietly in a suitable roosting spot while it awaits cues favorable for actual departure on migratory flight. This period normally lasts for a few hours until nightfall (for nocturnal migrants, which include the majority of songbirds), when the bird will take off. If conditions are unfavorable, it will sleep through the night and begin the next day by re-entering the "Feeding State" state, as shown in the figure. If the bird is over terrestrial habitats at dawn, after its overnight flight, it will normally land and begin searching for suitable habitat in which to rebuild its fat reserves. Once such habitat is found, it will reenter the "Feeding State." The sequence of events is repeated daily until the bird reaches its destination.

When I originally developed and published this sequence back in 1976, based on our waterthrush experiments and the work done by Donald Farner and James King and their students with caged birds, I thought that most insectivorous songbird transients would have to establish feeding territories, like those we found in the waterthrush, in order to sequester sufficient food to fatten at stopover areas. Subsequent research by a number of workers, especially Frank Moore and his students at the University of Southern Mississippi, has demonstrated that food resources in the form of arboreal and volant insects, at least in spring, are so abundant along major migration routes that territories seldom are necessary (an observation that will come as no surprise to anyone who has had the opportunity to take a passage through any bayou or salt marsh during these times).

The Northern Waterthrush uses different foods, however. It depends on soil invertebrates in muddy borders—a habitat sufficiently restricted in distribution at Welder along the Texas Gulf coast during the spring of 1975 so as to require territory establishment by waterthrushes wishing to rebuild fat reserves during stopover. Thus, my observations, which, following Occam's razor, I took to be representative for stopover behavior for all songbird migrants, were, in fact, applicable only to waterthrushes.

Nevertheless, although territorial behavior is not always necessary, most if not all birds likely follow the sequence of fattening-flight-stopover-fattening laid out in Figure 5.3. It is likely that the interplay between the environment, the bird's physiological needs, and the behaviors necessary to meet those needs are under tight hormonal control, but at present we have only the vaguest notion for what precisely constitutes those controls.

The experiments performed at Welder provide a remarkable insight into the behavior and physiology of free-flying transient birds, and, to my knowledge, they have never been duplicated or even approached. They set the stage for what should have been follow-up experiments in the lab where details of hormonal and other physiological changes, and their associated behaviors, could be assessed, manipulated, and tested under controlled conditions. I tried for forty years to find the funds to do this work (seven proposals to the Smithsonian Scholarly Studies program, three to the National Science Foundation), but I was unable to do so. As a result, we still have a poor understanding of these linkages. Science, however, is patient. Eventually, someone will follow up, and the critical ties between behavior, physiology (hormones, fat deposition), and weather for transient birds will be clearly understood.

Arrival

The sequence of intensive feeding followed by long flights that constitutes the process of migration continues from the bird's departure from its wintering area until its arrival at its breeding area. Adult males generally return to their breeding territory of the previous year, if they had one, or at least to the immediate vicinity. Adult females likely do the same, although rates of return, as measured by recapture of birds banded the previous year, are often much lower for females than for males. Many researchers have assumed that lower return rates to breeding territories of the previous year mean lower female survivorship. However, as discussed in Chapter 6, data on Prairie Warblers indicates that female return rates likely are comparable to male return rates to the *vicinity* of where they had bred. The difference lies in the fact that many females choose to mate with a different male, using a territory different from that used the previous year.

There is similar confusion regarding whether juvenile birds migrate to the areas in which they were born and raised. Return rates based on banded bird recaptures often are low, on the order of one or two percent, for these second-year birds, regardless of sex. Nevertheless, juvenile return to place of birth is obviously the best strategy, and it is likely followed by most individuals. Low recapture rates have several possible explanations, including the following: *low survival rates*—most young of most songbird species die during their first year; *low probability of recapture*—if the young bird does return to its birth territory, it is not likely to be allowed to hang around very long if the territory is already occupied, in which case, the chances of recapturing it are low; *other options*—the fact that young birds appear to use part of the postbreeding period to explore the area around their birth territory may mean that they will return to a likely-looking spot discovered during this exploration rather than the territory on which they were hatched. Studies of European Pied Flycatchers in Finland have demonstrated that when extra territories are created through the process of installing additional nest boxes in the forest, these territories often are taken by young hatched on neighboring territories the previous year.

CHAPTER 6

BREEDING
Period

La Belle Dame sans Merci.

—John Keats, "La Belle Dame sans Merci: A Ballad"

The breeding period has been, and remains, the traditional focus for understanding life history for both migratory and nonmigratory species alike. One only need examine the many summary volumes written both for professional ornithologists and lay people to grasp this critical fact. Consider the lifetime best seller and archetype of the genre *A Natural History of American Birds of Eastern and Central North America* by Edward Howe Forbush and John Bichard May, illustrated by Louis Agassiz Fuertes, Allan Brooks, and Roger Tory Peterson. There are 418 species treated, mainly from eastern and central North America, over half of which are migrants that spend four months or less on their breeding grounds. The only mention of the wintering ground is under "Range." Even the exhaustive 716 accounts for every North American bird authored by the world authority on each species under the auspices of the American Ornithologists' Union and the Academy of Natural Sciences gives the wintering period short shrift for most species, and distribution maps did not even show wintering range outside of the continent for long-distance migrants (this omission has been corrected to a degree in the new version of this publication, now titled *Birds of the World* and published online by the Cornell Laboratory of Ornithology).

It could be argued that such treatment emanates from a lack of information, but this excuse simply begs the question, if a book purports to treat the life history of a species, how can the majority of its annual cycle be ignored? The answer is simple. The missing portion is not deemed to be important. Stripped to its essence, this is the argument behind the Northern Home Theory. The migrant niche exists only on the breeding ground. Once it leaves this area, it is a wanderer—a stateless being driven from pillar to post by resident ecological counterparts, subsisting on ephemeral resource superabundances from the time it departs the breeding ground until its return the following year. Although a great deal of outstanding work has been done over the past twenty or thirty years to rectify this problem, the fact remains that the Northern Home Theory, and its relative, the Migratory Syndrome, assume an evolutionary focus on the breeding area.

My thesis research in Minnesota, Texas, and Mexico focused on the investigation and testing of this theory. Very early on in our work, we found evidence to indicate that migrants were not, in fact, homeless visitors to the tropics; that, indeed, they were quite literally "home." Years of subsequent research (by me and others) have confirmed this fact. The wintering period is critically important. But it doesn't stop there. If migrants have niches on the wintering ground, what about the breeding ground? The possibility of wintering-ground origin for migrants inverts the entire concept.

The idea that most Nearctic migrants to the tropics appear to have evolved in the tropics poses an apparent fallacy: how can migrants have evolved to occupy niches in both a specific tropical environment and a specific Nearctic environment? The answer, in my opinion, is that they cannot. So how do we explain this conundrum? I suggest three possibilities which are not mutually exclusive: the first is that food is superabundant during the summer periods in the Nearctic, relaxing the need for specialized feeding structures and behaviors; second, the number of competitors for that food is greatly reduced in comparison with tropical environments, further relaxing niche strictures; and, third, there may be structural similarities for certain microhabitats between some tropical and Nearctic habitats that allow seasonal exploitation in the latter. These hypotheses invite a closer look at the migrant breeding period from an altered perspective.

For one thing, it means that survival is taken out of the equation as a reason for migrating north in spring. If survival is all you are worried about, then you should stay on the wintering ground or somewhere in between the wintering and breeding ground—wherever food is most abundant and easily and safely harvested, which is what many second- or third-year birds of long-lived species, such as some hawks, sandpipers, and seabirds, often do. Migration to the breeding ground is all about sex, and if you can't compete, then you might as well keep your powder dry and hang out in the best place you can find for survival until you are old enough and experienced enough that you can compete, at least if you are a male. If you are a female, then the equation is quite different. You don't have to compete for the most part. In fact, *you* are what *they* are competing for.

There is an insightful cartoon by Jennifer Berman depicting the thought frequencies of men and women as pie charts. For women, seventy-five percent of the chart is devoted to "The Relationship," with "sex" as a small subset of this section; the other twenty-five percent of the female chart is divided between food, aging, pets, men trashing, and having to pee. For men, about ninety percent of the chart is divided into three equal portions: sex, career, and sports (with "The Relationship" as a small subset of the "sex" section). The remaining ten percent is divided into aging, going bald, and strange ear and nose hair growth.

Now you may cringe a bit at my seemingly gratuitous incorporation of gender stereotypes, but I do have a point. We all have our own opinions concerning the accuracy of Berman's observations based on our own experiences, but for the purposes of this discourse, they are irrelevant. Most people would at least agree that Berman captures some elements of truth. Sharp motivational differences exist for male and female migratory birds as well, but they are quite different from those of humans. For one thing, male focus on "The Relationship," in terms of time spent, is much greater than that of females in migrant birds. Also, of course, reproductive activities are seasonal. For female songbird migrants, "The Relationship" does not play any role at all in their lives until their actual arrival on the breeding ground and their selection of a male and his territory. My hope is that leading off our discussion by highlighting these differences will help the reader to understand that it is *useless* to consider avian sexual relationships through a human lens.

Most people, including nearly all scientists, would scoff at the very idea that they would allow such a ridiculous distortion to color their views. Mr. and Mrs. Scarlet Tanager clearly have quite a different partnership from Mr. and Mrs. Smith. And yet, a cursory review of the avian mating system literature should convince most skeptics that male dominance over their mates has been, and remains, the guiding principle for understanding how bird sexes relate. The story that I will present to you takes a very different view.

Consider the data from Val Nolan's superb study of the Prairie Warbler. Female breeding-ground arrival does not occur until toward the end of April in Indiana, and that's when "The Relationship" begins. By the end of July, breeding is complete for most females, and "The Relationship" is over. She is on her own until the following April. Thus, for nearly nine months of the year, the sole focus of female behavior is on personal survival. It is only for a short three months in summer that her motivations shift from survival to a balance between survival and reproduction. Even during this period, however, male motivation is basically under her control. Male "dominance" of the female can play no part. This situation stems from the fact that successful copulation requires complete female cooperation for nearly all bird species because males have no penis, with a few exceptions including some ratites (such as ostriches) and waterfowl. Sperm is transferred when the cloaca (external opening for both the alimentary and reproductive tracts) of the male is placed against that of the female. For this to happen, the female has to move her tail aside, remain relatively still, and relax her cloacal sphincter. Thus, females shape most male reproductive behavior in birds. No rough stuff, or at least none that is not female-sanctioned (more on this later).

FIG. 6.1. Male (*left*) and female (*right*) Northern Pintails. PHOTO BY J. M. GARG, CREATIVE COMMONS ATTRIBUTION LICENSE

Females often are held responsible for male differences in size and ornamentation. According to this hypothesis, females choose larger, more aggressive and brilliantly colored songster males over their less-endowed counterparts, hence favoring the evolution of these behaviors and structures through sexual selection. This explanation may be correct, but research on the topic raises some doubts.

An experiment conducted by Lisa Sorenson with Smithsonian scientist Scott Derrickson during Lisa's postdoctoral work at the Conservation and Research Center (CRC) in Front Royal, Virginia, in the 1990s is an example. She wanted to assess the value of male size and ornamentation in Northern Pintails in pair formation by presenting two males differing in these characteristics to a female and then observing which male the female chose as her mate (Figure 6.1). Many trials of the experiment were performed, but the expected result did not occur—females chose the "lesser endowed" male about as often as the one "more greatly endowed." If male adornment is a key factor in female mate choice, then the results make no sense. However, there is another hypothesis that could explain the outcome. What if male size and adornment serve mainly in competition with other males? If that is so, then the experimental setup of two males penned together on a small pond into which the female is introduced may just be weird for the female. In the wild, one of the males, probably the larger, more ornamented male, will have driven the lesser male away; but in the CRC pens, that can't happen. Regardless of whether or not this alternative explanation is correct, the experiments clearly demonstrate that male size and ornamentation are not especially important for female pintails in terms of mate choice.

The same may be true in migratory songbirds. In the vast majority of birds, males and females are alike in appearance, at least for those that are principally monogamous. However, in about a third of migratory species, males, on average, are more brightly colored than females. Most researchers have assumed that male ornamentation in these species was an evolutionary result of female choice. However, experiments with Red-winged Blackbirds by F. W. Peek raise questions concerning this idea. Peek darkened the normally brilliant scarlet epaulets of male redwings and found that these birds were unable to maintain their territories in competition with other males (Figure 6.2). This result suggests that ornamentation, at least in this migrant, may result more from intramale competition than female choice. But if that is so, what factors might be important to returning migrant females in terms of a mate?

The problem confronting the female when she first arrives back from the wintering ground is that she doesn't have a lot of time during which to make the most important decision of her life—who is going to sire and help raise her offspring? Much of her decision will focus on the excellence of the territory that the male has managed to stake out and defend (as mentioned before, males arrive a week to ten days before females on the breeding ground).

"Quality" in a mate is, of course, critically important. Yet for a newly arrived migrant songbird female, an essential part of that quality is the territory: its size, habitat value in terms of provision potential for offspring,

FIG. 6.2. Male Red-winged Blackbird in territorial display. PHOTO BY LINDA PETERSEN

and number and worth of potential nesting sites. Considered from this perspective, song and appearance may serve principally for the female to confirm that the male is a member of the correct species, subspecies, or metapopulation. The relative size, aggressiveness, singing persistence, and plumage brilliance of the male may, in fact, be more crucial in terms of competition with other males for territories in monogamous species than in attracting females. In other words, it may be that intrasexual selection (resulting from male-male competition for territory) is more significant in the evolution of these characteristics then intersexual selection (female mate choice), at least for males.

The normal arrival and territory-occupancy pattern for songbird migrants returning to the breeding area from their tropical and subtropical winter homes lends support for the second of these hypotheses. Males of these species generally arrive back on the breeding ground a week or two earlier than females (protandry). If they are older birds that have bred previously, there is a high probability that they will attempt to occupy the same territory that they held in earlier years. On arrival, they will try to defend as large an area as they can against their neighbors or young males attempting to locate and defend a breeding territory for the first time.

Older females also come back with high regularity to the area where they bred the previous year, but they often do not choose the same male or territory. Val Nolan notes for his Prairie Warbler population in central Indiana, "Whereas probably all surviving males homed to the territory of the preceding year, a large number of females did not." A number of researchers with other migrant species have found the same thing, and several have assumed that this meant simply that females had a much lower survival rate. Nolan's work documented that this assumption was false, at least for Prairie Warblers. He found that older females returned to his study area at similar rates as older males, but they often chose to settle on different territories (with different mates). This behavior makes a great deal of sense if the female's focus is on locating the highest quality territory possible, rather than some aspects of the mate.

The importance of number of potential nesting sites as a possible factor in terms of territory quality has been demonstrated quite clearly by work with European Pied Flycatchers. This bird is a long-distance migrant that breeds in western Eurasia and winters in the African tropics. It is a cavity-nester and studies in a Finland population demonstrated that numbers of breeding territories in a given forest plot could be increased simply by adding nest boxes. This work also emphasizes the complexity of the assumed relationship between breeding territory and the food needs of the family during the breeding season. If the number of breeding pairs of migrants in an area can be increased simply by adding nest sites, what does this mean in terms of the importance of food with regard to territory quality and breeding success? That there is some positive association between food availability and reproductive success is demonstrated by numerous studies documenting a correlation between increased clutch size in accordance with increased food availability. Nevertheless, the relationship is not a straightforward one.

The Northern Home Theory holds that migratory birds evolved on their breeding grounds, their foraging behaviors and structures being derived from this process. Thus, according to this theory, breeding habitat is presumed to be critical to reproductive success, and lack of it dooms the species to extinction. The critical unit for an adult individual is the breeding territory, which, for most songbird migrants, consists of a plot of ground in appropriate habitat that provides the breeding pair with all the necessary elements for successful raising of a brood: food, nest site, and shelter. This is often termed the "Type A" territory, following the classification system of Margaret Morse Nice. By definition, the Type A breeding territory is exclusive by nature, that is, the male bird establishes and defends the territory, preventing entry by other males of the same species.

The timing and dynamics of this process can be summarized as follows: the male arrives on territory in spring, defending the territory boundaries by singing and by attacking, fighting, or chasing male competitors. When the female arrives some days later, she selects a mate/territory, courtship begins, followed by nest construction, copulation, egg-laying, and incubation (which is the sole duty of the female in most songbirds). When the young hatch, usually in twelve to fourteen days for small songbirds, both parents feed them until fledging in ten to twelve days. Post-fledging, the young continue to be fed by the parents for another three weeks or so until they become independent. In some species more than one brood is raised, in which case, the adults begin the process again. In any event, whether single or multiple-brooded, once the last brood has achieved independence, the adults begin molt, a process that takes one to two months; when this is complete, they begin preparing for migration, laying down fat reserves. As soon as reserves are sufficient, they depart southward on their journey to their winter quarters.

There are as many variations on this theme as there are migrant species, but the above describes the basic elements for many migratory songbirds. For me, the first inkling that this existing paradigm of migrant breeding ecology was flawed came with the discovery in our Veracruz work that several species of migrants occupied territories in primary (never previously cut) tropical forest throughout the winter, even returning to them in subsequent winters. This finding suggested that these birds had niches in these forests, that is, that they occupied an ecological space that was all their own. But for this to be true, they had to have foraging behaviors and structures adapted to exploit the resources of the tropical forest environment. In other words, they had to have evolved in that environment.

The logic of this argument is inescapable, but it was not until a few years later, when compiling my review of literature on Nearctic avian migrants in the neotropics for the World Wildlife Fund and US Fish and Wildlife Service, that I learned about a huge piece of corroborating evidence: nearly half (forty-eight percent) of Nearctic migrants have breeding populations in the neotropics. That is, 159 species of migrants have one population or more *resident* in the tropics year-round, while others migrate to breed in the Nearctic temperate or boreal zones. Examples include the Yellow Warbler, Great Blue Heron, Eastern Towhee, and Red-eyed Vireo, among many, many others.

Field Work on the Migrant Breeding Territory

An additional key element undermining the received wisdom of breeding territory centrality to migrant life history came from some work done during the University of Minnesota's field biology summer session at the Lake Itasca campus in 1974. My wife, Bonnie, was not excited about living in tents for two months (as we had the previous summer), so after leaving Texas in the spring, she and the kids headed east to stay with my mother while I went to Itasca for the summer semester. My PhD advisor, Dwain Warner, had taught field ornithology there for thirty years, but for some reason the university hired Dr. Andrew Berger to teach the course that summer. In any event, I was hired to be his teaching assistant.

I knew Andy by reputation as coauthor, with Josselyn Van Tyne, of the wonderful *Fundamentals of Ornithology*, the standard text on the subject for many years. My dad had given me a copy when I was twelve, on the occasion of my promotion to First Class Scout (he was an Eagle Scout, but I washed out after Star to pursue other interests). The preface to that book tells you a lot about Andy Berger: gentleman, scholar (principally as an anatomist), and loyal collaborator with the renowned Van Tyne, who had passed away during the project. Andy was a terrific person and superb field biologist (discoverer of the first nests known to science of several Hawaiian species), however, at nearly sixty years of age and having spent the past decade in Hawaii, he was not interested in the intense schedule and long hours involved in presentation of a field course in northern Minnesota, so he pretty much turned responsibility for it over to me.

Mario Ramos, my colleague from thesis work in Veracruz, was there to help, and we decided to set up a bunch of mist nets at LaSalle Creek Bog to capture and band birds as part of the training for our students. This little project was my introduction to the migrant breeding period, and eventually it became the first of the three chapters of my doctoral thesis (I. Breeding, II. Migration, III. Wintering). With the assistance of a couple of our most avid course participants, John Barber and Judy Wilson, Mario and I ran twelve mist nets fourteen hours a day for three weeks, accumulating about 3,500 net-hours during the last week in June and first two weeks in July.

What we found fascinated me. Eight species had breeding territories on the plot, as evidenced by observation of color-banded birds, recapture, and nest discovery. The site was small, a couple of acres, so there appeared to be room for only one or, at most, parts of two territories for each of the species. We confirmed this supposition by observing response of color-banded territorial males using playback of song. Nevertheless, we captured males in our nets far in excess of the number of territories. For instance, we knew from playback that a single Chestnut-sided Warbler pair had a territory on the site, and we knew where the nest for this pair was located. Yet over the course of our netting period, during the height of the breeding season, we captured fourteen males. Most of these birds were never seen again after initial capture. The territory owner died on July 7. At the time, his mate was incubating a clutch of three eggs. By July 11, an unbanded male had taken over defense of the territory (and presumptive cohabitation with the female), continuing there, singing, every day until we left.

There were three noteworthy aspects to these findings. First, the breeding territory for small passerines, which was defined as "exclusive" by most experts, was not exclusive. In fact, there were "floater" males moving through the territory continuously for all eight of the species known to have territories on the site (many other studies have found this to be true, including one by Darwin). Second, if a territory owner ceased his defense (display and vocalization) for any reason, one of these floater males would take over the territory in short order. Third, female floaters were scarce or nonexistent.

The data from this little study provided clear evidence, for me at least, that food was not a critical aspect of the migrant breeding territory, and that females were the arbiters of breeding-period social organization. In fact, the contrast for the species that I observed between winter territory (defended against conspecific intrusion regardless of sex) and breeding territory (basically defended only against males perceived as rivals—at least once mates were incubating) could hardly be more stark.

One of my graduate students during my tenure at Texas A&I University (now Texas A&M University, Kingsville), Dave Swanson, and I discovered further evidence of the disconnect between food requirements and breeding territory in our studies of the White-winged Dove, a migratory game species that has breeding populations in South Texas that winter in the dry Pacific lowlands of Central America. We found that, while male whitewings defended breeding territories in thorn forest, both they and their mates did all of their feeding in neighboring sorghum fields. The breeding territory provided no food whatsoever.

Joe Meyers and his students found a similar situation in Painted Buntings on Georgia barrier islands. Breeding territories for nesting pairs were located in oak forest, but the birds left the territory to feed in nearby marshland.

Hypothetical Purpose
of the Migrant Breeding Territory

So, if food is not the critical issue in terms of breeding territory establishment and defense (in my opinion), then what is? Breeding male behavior (song, territory, and so forth) is all about affairs of the heart, or, more crudely, sex. Males attempt to defend territories that provide the highest number of excellent nest sites possible with feeding areas in the immediate vicinity. Females arrive later than males to allow time for males to establish the best quality territory they can. A female can use this indicator in her choice of a mate. Thus, I suggest that sexual selection, not food, is the driver behind breeding territory size and structure, depending on the ecology and life history of the species.

MacArthur's brilliant postdoctoral research fellow Steve Fretwell, in his ground-breaking work *Populations in a Seasonal Environment*, was among the first to put forward the heretical notion that food was not limiting for migrants during the breeding period, and that, in fact, "there is no a priori reason to expect differences between coexisting breeding bird species, which may be as similar as they like [since food is superabundant]." In essence, what Fretwell says is that looking for differences in foraging behavior among species on the breeding grounds (which was a major activity of field biology graduate students for a generation—including MacArthur himself) is sort of like searching for such differences among species visiting a bird feeder. The dissimilarities observed are irrelevant because there is plenty of food for all, regardless of what method one uses to harvest it. He further proposes that is not the case on the wintering ground, where food often *is* limiting and behavior and morphological differences employed in resource defense and harvest are critically important to survival.

For me, these ideas help to explain a curious fact discovered during preparation of my book *The Ecology of Migrant Birds*, to wit, of the forty-eight migrant species that breed in the *forests* of northern Europe, all but thirteen winter in *open habitats* of tropical Africa (savanna, scrub, desert, grassland). Assuming that these birds evolved in these tropical non-forest habitats, one must conclude that their ability to use temperate zone forest for breeding has to do with the fact that food is superabundant there during the summer season so they don't have to have special adaptations to compete for these resources.

The size and structure of the breeding territory is determined largely by the kinds of nesting sites needed and the type and distribution of food resources required by members of migrant species to feed themselves and their young. If the food and nesting sites are to be found on the same piece of ground, as is the case for most migrant songbirds, then the Type A territory covering a defined defended area, where most feeding, copulating, and nesting take place, is the result, which is what most members of these species defend. Where feeding and nesting areas are separate, as is the case for most waterbirds, swallows, and the like, the breeding territory can be quite small, often limited to the nesting site alone.

Mating system structure is also related to food and nesting site distribution, but filtered by female fitness needs. As discussed above, lack of male ability to force females to mate means that females hold the ace when it comes to shaping male reproductive behavior and, of course, the type of mating system appropriate for producing the largest number of offspring during her life. For migrant songbirds, seasonal monogamy is the most common mating system structure. The assumed reason for this is that songbird young are born naked and helpless, requiring intensive care in terms of incubation and feeding for a couple of weeks before fledging, and additional care and training for some time after fledging. Without male assistance, the number of young that could be raised to independence would be far fewer, dramatically lowering fitness for both members of the pair. Male participation in this scheme is more or less forced on them by circumstances. Females likely choose a mate based mostly on the quality of breeding territory. If the male should leave after copulation, he would not only decrease the number of his offspring likely to be raised, he would also lose his territory, as well as any chance of further copulation with that female. Since nest failure due to predation runs fifty percent or more, this option is a poor fitness choice. Also, the number of other females that would allow copulation with an itinerant male are likely to be limited.

As we found in LaSalle Creek Bog study, there are few or no female floaters. The presumed reason for this is that if a female cannot find a territorial male with which to pair, she will probably settle on the territory of an existing pair, accepting a polygynous relationship as preferable to not breeding at all. Rates of what is known as "serial polygyny" vary according to environmental circumstances, but they often run at ten percent or more of the breeding population among migrant songbirds.

Potential Consequences of Female Choice

Popular ideas regarding the lives of birds are built upon misconceptions and misunderstandings, perhaps none more so than those concerning avian mating systems. It is true that many migratory birds are monogamous during the breeding season, but it is important to remember the key that favors such a system, namely that the members of the pair can contribute more offspring to the next generation by cooperating in raising the young to independence than they could by following some other system. If circumstances change in terms of what constitutes male quality, then structure of the mating system can be expected to change rather quickly as well.

An example of such a change may be the discovery of surprising rates of extrapair paternity (adultery!) in many species of migrant songbirds. Since the 1990s, advances in genetic techniques have made it possible to determine the father of offspring produced on a given territory in purportedly monogamous situations. This information has been shocking, revealing rates of young for which the male territory owner (and caregiver) is not the parent as high as fifty percent or more in some songbird migrant populations. Usually, the actual biological father turns out to be a male, often a year or more older than the putative mate, from a neighboring territory. This circumstance is referred to as "extrapair copulation," "extrapair fertilization," or "extrapair paternity" in the literature. What makes this finding so astonishing, for me at least, is that the assumed fitness payoff for territory defense for a male is greatly diminished if a significant portion of the offspring raised are not his own. Interestingly, rates of extrapair copulation are far lower or zero in tropical resident breeding pairs of the same or related species.

It is my belief that high rates of extrapair copulation in migrant songbirds represent a relatively recent phenomenon, perhaps resulting from the fact that wintering habitat destruction reduces the number of experienced males returning to breed and brings more young males into the breeding

FIG. 6.3. Female Greater Prairie-Chicken (*second from left*) visiting a lek. PHOTO BY CELESTYN BROZEK

population, thereby favoring the female search for higher-quality males to sire their offspring. Examination of paternity patterns in museum specimens taken a century or so ago could test this idea, but this has not been done to date so far as I am aware.

Regardless of whether or not frequent extrapair copulation is a new development in songbird migrant breeding behavior, it demonstrates the degree to which females control male reproductive success, as does the variety and structure of different mating systems. When female need for male help in raising young is diminished, as is the case for many species with precocial offspring (those that hatch able to forage on their own), mate quality in terms of competition with other males assumes a much greater value, resulting in various levels of polygyny or even a lek system, as seen in the Greater Prairie-Chicken, where female contact with adult males is limited to a brief visit to male group display areas (leks) solely to select and copulate with the dominant male.

These visits are pretty comical to the disinterested observer. In the Greater Prairie- Chicken, leks are usually pieces of ground nearly devoid of vegetation and surrounded by open grassland. Perhaps eight or ten male birds defend small bits of this ground, a few square feet, where they perform extraordinary dances and vocal displays each day during the breeding season, from sunrise until midmorning. Females visit these leks, strolling nonchalantly, and seemingly indifferently, among the furiously capering males, finally choosing one with which to mate (Figure 6.3). Nearly all of the visiting females choose to mate with the male having the most-central territory, ignoring all of his dancing buddies.

In my opinion, our understanding of the level of avian female power over male fitness is a fairly new idea. Personally, I would date it to the publication of *Sociobiology*, E. O. Wilson's revolutionary work. Therein, Wilson pulled together a vast amount of information from across the entire field of zoology to put forward what he called the Modern Synthesis, in which "each phenomenon is weighed for its adaptive significance and then related to population genetics." Of course, this statement sounds like Darwin's old synthesis. The difference, as made clear in the text, is that the term "each phenomenon" refers to behavior as well as every other aspect of species structure and function. Wilson's summary made clear his belief that the *actions* of each individual, regardless of taxonomic group, were shaped by natural selection.

For most *Homo sapiens* this thinking is counterintuitive, particularly when applied to our own species, as Wilson did in Chapter 27 of his book. In fact, many people found it to be infuriating, propelling the work well beyond the level of an important biological treatise to cultural zeitgeist directly relevant to heated debates concerning women's rights and other aspects of human relations.

Whether or not one accepts all of Wilson's ideas as they apply to us, ignoring their importance for other animals, became awkward in biological circles. Nevertheless, biologists are people too, and it has taken awhile to fully appreciate what it means to free the female bird conceptually from a male-dominated system to one in which reproductive behaviors are considered solely in terms of cost-benefit fitness analysis for the individual rather than a particular moral code, the lens through which avian mating systems were viewed by many until quite recently.

To illustrate my point, I will relate a story. Back in the mid-1970s, I took a group of students to visit a large cattail marsh in southern Minnesota. While there, I observed a curious sequence of behaviors involving Red-winged Blackbirds. Hundreds of redwings had nesting territories there, and every once in a while a female would fly up, presumably from her nesting area, and out across the marsh. Often, a male redwing would take off in evident pursuit. Usually when this occurred, the female, which is smaller but more agile in flight, would suddenly change direction, immediately losing her pursuer. However, on occasion a female would continue in straight flight until three or four males were in chase and then dive out of sight into the cattails to be followed by all of her swains. What happened then must be left to the imagination, but in a minute or two all participants would rise up and head off to their respective starting points.

One could consider these encounters as examples of forced copulation. However, one could also think of them as testing of potential mates, particularly since it appeared that the female could avoid encounters by erratic flight maneuvers or by not flying across several male territories, and, even if forced down by the group, she could refuse to move her tail out of the way or relax her cloacal sphincter.

I wrote these observations up into a short note that I planned to try to have published in the leading bird journal *The Auk*. I wanted a tough, skeptical examination before submitting the piece, so I asked Harrison (Bud) Tordoff, director of the University of Minnesota's Bell Museum of Natural History at the time, if he would provide a critique. He did, and when he handed it back said that he had discussed the paper with another Bell Museum professor, Frank McKinney, who was working on a major review of forced copulation in waterfowl. McKinney, a former student of Nobel laureate Niko Tinbergen, had advised Bud that my little commentary would contribute nothing not covered in greater detail and cogency in his planned major review of the subject for the journal *Behaviour* (with coauthors Pierre Mineau and Scott Derrickson), and that I should not bother to try to publish it. Tordoff made clear to me that he agreed with this assessment. So I simply shoved the account into a file and forgot about it.

When McKinney's article came out, I realized I had been misled. To explain how, I will need to provide a little background on breeding biology of migrant ducks. Pairing for many of these species occurs on the wintering ground, where females choose a partner from among the many males present. On departure on spring migration, the male accompanies his mate back to where she bred or was hatched the previous year. On arrival at her breeding pond or lake, the female builds the nest, copulates with her consort, lays her eggs, and commences incubation. Once she is incubating, the male's job as part of a mated pair is usually complete and his association with his mate is thus over for the year, at which time he will often depart for another body of water separate from where his family is located. Once this has occurred, the female is on her own. The ducklings, which are precocial, are her sole responsibility. During this period, if she should happen to fly away from her breeding area, she will often be chased by several males who will actually compel her to the ground and copulate with her. They are physically able to do this because a male duck has an internal penis that he can evert and force into the female's cloaca.

Based on McKinney's assurance, I believed that he would address the costs and benefits, in terms of fitness, of these behaviors from both the male and female perspectives. After all, as a member of my PhD committee, it was McKinney who told me to prepare for my final oral thesis defense by reading Wilson's *Sociobiology*. However, he did not address the phenomenon from the perspective of both sexes in his paper. Instead, his entire focus was on the male in terms of behavior, physical attributes, and fitness benefits. In other words, he began his consideration under the assumption that the activity amounted to forced copulation and that the female role was simply as a recipient of a potentially dangerous activity engaged in for the sole benefit, in a fitness sense, of the male. The fact is, though, that we do not know that this assumption is correct. Removing or replacing this assumption allows us to examine the behavior in a different light. What if it is the female who is in control of the situation?

Looked at from this angle, one has to ask, just as with the female redwing, why was she flying high in the first place? The body of water where she is raising her young presumably has everything she needs to accomplish this purpose, which is why she chose it. Why should she leave it? Well, there is one obvious thing that it doesn't have that is absolutely critical for her fitness, namely, a male. What if the clutch was destroyed or the offspring eaten after her mate's departure? It's true that ducks can store sperm for later use, but maybe that is not the best fitness outcome for the female. Maybe there are advantages to testing males head-to-head and allowing the winner to copulate. Viewed from this angle, the anatid penis may be more of an adaptation to allow completion of copulation when in the middle of a hurly-burly rather than a means of forcing the female, who might in fact be complicit.

My aim in recounting this episode is to emphasize how preconceptions can influence our analyses of observations, causing us to make unwarranted assumptions. I do not say that my interpretations are true—only that they might be. Rather than assuming that they are not, they should be tested. Some of my readers may find this imputation of the female duck character disgusting, but again, I reiterate, I have a point. The behavior of each individual bird presumably is crafted solely for the purpose of contributing as many offspring as possible to the next breeding generation. We have to put aside our human perspectives in evaluating avian motivation. Birds, so far as we know, have no moral compass, and therefore their behaviors cannot be judged on this basis.

One final note on the topic. If, as I claim, female migrants exert such control over male fitness, then sexual selection in the form of female choice may have extraordinary power in terms of producing rapid change in male morphology and behavior, a possibility I will discuss in greater detail in Chapter 9.

CHAPTER 7

POSTBREEDING
Period

And he vanished out of their sight.
—Holy Bible, Luke, 24:31 (King James Version)

A curious thing happens to many species of migrants during the middle of the supposed temperate zone breeding season. They disappear! Woodlands filled with song in May and June become silent later in the season. Where'd they go? The purpose of the breeding period is, *helllooo*, breeding! But what happens to the pair and their offspring when breeding is completed, after the last chick has reached independence, usually by early to mid-July for most songbirds? The general assumption has been that they just sort of loiter on the breeding territory until departure on fall migration. One need only check out the hundreds of species accounts for migrants in Arthur Cleveland Bent's *Life Histories* or Poole's *The Birds of North America* or Del Hoyo et al.'s *Handbook of the Birds of the World*, or any other such compilation, to verify the veracity of this claim. Indeed, all of the hundreds of summaries of avian life history data of which I am aware report *three* phases for a migrant's annual cycle: breeding, migration, and wintering. But they are wrong. There are at least *five* phases for many migrants: breeding, postbreeding, fall migration, wintering, and spring migration, each of which has its own particular structure, timing, habitat requirements, population dynamics, and evolutionary strictures peculiar to each species.

Among these phases, the postbreeding period is particularly myste-rious because its timing can vary so much from individual to individual, even for members of the same species living in the same place. Neverthe-less, it is important that we recognize it for what it is because, once the postbreeding season begins, each individual is no longer a member of a family unit; they are on their own with their own needs, in terms of food, habitat, and competition for resources, in ways that are quite different from the needs of the breeding season. Reproduction is no longer the focus. For adults, this represents a major shift in motivation from raising as many offspring as possible to maximizing one's personal survival.

Earlier students of avian breeding season life history assumed that what happened when the forest fell silent was simply a continuation of the "breeding period," lasting until the birds left on fall migration. While it was well known that migratory ducks, particularly males, migrated away from breeding ponds to special molting sites in the middle of the summer, the fact that individuals of many other migrant species disappeared from for-est breeding habitat, months before appearing south of the breeding range on migration, was hardly recognized.

Field Work on the Migrant Postbreeding Period

The possible existence of this portion of the migrant life cycle came to my attention more or less by chance. Kevin Ballard was a young graduate stu-dent of mine at the University of Georgia during my time there. In the summer of 1980, we set up several mist nets in order to capture and band birds on a small plot of ground in Athens belonging to the university called Horseshoe Bend. The purpose of the work was mainly to give Kevin and another graduate student of mine, Carol Gobar, a chance to get some expe-rience in conducting a field study.

The site was about thirty acres, consisting of a peninsula bordered on three sides by the Oconee River and on the fourth by a four-lane highway. Two habitats occupied the peninsula: riparian forest bordering the river and scrubby field in the interior. The forest consisted of a strip one hundred to three hundred feet in width in which American elm and red maple were the dominant tree species. Forest canopy height was fifty to seventy feet. The "old field" habitat consisted of dense herbaceous undergrowth of aster, ambrosia, caster, and so forth with emergent saplings, six to ten feet in height. The nearest similar old field habitat was five hundred feet away.

Sixteen mist nets were placed roughly one hundred feet apart along a path that ran through the forest in mid-April. On July 16, we placed an additional seven nets in the old field. Nets were open from daybreak until noon, and the sampling continued from April 17 until October 15, 1980. A US Fish and Wildlife Service band was attached to each captive, and the following data were recorded: date and time of capture, net number (habitat), species, sex, age, mass, and amount of subcutaneous fat and molt. We captured 374 birds of 55 different species during 4,733 net-hours. Based on individuals heard singing, nests observed, or observations of young being fed, in addition to capture data, we concluded that a minimum of seventeen species bred on the Horseshoe Bend study site in 1980. Another twenty-one species were captured in spring (April-May) or fall (September-October). We presumed that these birds were transients moving between breeding and wintering areas.

A third group of species does not fit into either of these categories. These birds are known to breed in the state of Georgia, but there was no evidence of their presence on the old field portion of the site until they were captured there in July and August. Some of these species are known to breed commonly in the Athens area, but for others there are few or no records as breeding birds for the region. Thirty-three individuals of fourteen species exemplify postbreeding movement away from the place where they had bred. When captured at our old field site, they were either out of their normal habitat, out of their breeding range, or not known to have been nesting on our specific site based on netting and observations.

We distinguished four subgroups based on habitat preferences, breeding range, and seasonal status in Athens:

1. *Species that bred on the site but not in the old field portion.* The Wood Thrush, the only bird in this category, was a common

breeding species in the riparian forest portion of the study area where singing individuals were heard from the beginning of the study (April 17) until July 15. Although thirty-three Wood Thrushes were captured in the forest, no Wood Thrush was seen, heard, or captured in old field habitat until July 21. Thereafter, four of these birds were captured in old field habitat during July and August. Two of the four individuals showed moderate or heavy molt, little or no fat, and had been eating pokeberries. No Wood Thrushes were captured in riparian forest from July 30 until September 17.

2. *Species breeding in the Athens area in old field habitat but not found as breeders on our study site.* Five species met these criteria: the Yellow-billed Cuckoo, Ruby-throated Hummingbird, Brown Thrasher, Prairie Warbler, and Blue Grosbeak.

3. *Species breeding in the Athens area but not in old-field-type habitats.* Two species met these criteria: the Acadian Flycatcher (riparian forest) and Pine Warbler (mature pine forest).

4. *Species breeding commonly in the north Georgia mountains that are rare or absent as breeders in the Athens area.* Six species met these criteria: the Northern Parula, Black-and-white Warbler, American Redstart, Ovenbird, Canada Warbler, and Scarlet Tanager. Note that the nearest Appalachian highland breeding habitat for these birds is located sixty miles to the north of the study site.

The results of this work indicated that most birds in our capture samples for July and August were neither in a reproductive nor a migratory state; in fact, it indicated that they were in a portion of the life cycle that was completely separate, having its own physiological, behavioral, and ecological requirements. Although postbreeding movement has often been equated with the beginning of fall migration, over half (nineteen of thirty-three) of the postbreeding wanderers we trapped showed some degree of molt. Though individuals of several species do overlap molt and migration somewhat, such birds usually represent a small percentage of the population based on our experience with thousands of birds captured as transients in Texas. Furthermore, few individuals of any species show heavy body molt during migration, a condition found in seven of the birds captured in the old field habitat (one Ruby-throated Hummingbird, three Wood Thrushes, two Northern Parula warblers, and one Ovenbird).

The lack of subcutaneous fat reserves also argues against these birds being in a migratory state. Only two of the thirty-three birds captured, a Canada Warbler and a Scarlet Tanager, showed more than light fat reserves. Also, five of the species (the Wood Thrush, Brown Thrasher, Ovenbird, Canada Warbler, and Scarlet Tanager) were captured at the site long before early migration dates for the species in north Florida, based on R. L. Crawford's twenty-five years of TV tower kill data.

A southward "drift" is indicated by the fact that several of the species captured breed commonly sixty miles or so north of the study area but are rare or absent as breeders in Athens or farther south based on data from T. D. Burleigh's *Georgia Birds*. To call this a migration, however, is not warranted, given the lack of fat reserves and the presence of molt in many cases.

Eleven of fourteen species in the sample of wanderers are single-brooded, long-distance migrants. The data suggest that these birds abandon their breeding territories shortly after completion of breeding and wander in search of abundant food resources that are available in many natural, terrestrial habitats of the southeast during the summer and early fall months. The species that did not disappear from our site were either permanent residents (the Tufted Titmouse, Carolina Wren, and Northern Cardinal) or double-brooded migrants (the Gray Catbird, White-eyed Vireo, and Common Yellowthroat).

The existence of postbreeding premigration wandering is supported by the fact that the birds neither sang nor gave call notes. These same species sing and call vigorously in defense of breeding territories, and they average one to two minutes per hour in defense of winter feeding territories based on our work in Veracruz. Even some transients defend feeding territories, using call notes and displays. The silence of these birds at our study site in

July and August, their lack of site tenacity, and their lax preferences with regard to habitat selection all indicate that food was not limiting, and that intra- and interspecific competition for resources were not particularly stringent during this postbreeding period.

Purposes of a Postbreeding Period

Our Georgia findings on this portion of the migrant life cycle were published back in 1987. Since then, several other papers have come out further documenting the movement of migrant species out of their forest breeding habitats into food-rich scrubby second growth, further establishing the timing of that movement (mainly July and August), and expanding the list of migrants known to do so to thirty-nine species. Not all species of migrants change habitats during the postbreeding period, but it is likely that all migrants undergo this portion of the life cycle, regardless of what habitats are used. This fact sort of begs a couple of questions:

> *What's the purpose of the postbreeding period?*
> *Why can't that purpose be satisfied within the breeding territory*
> *or habitat?*

One of the distinctive characteristics of the postbreeding period is that movements are quite different for members of different sex and age groups, as documented by two field studies—one of the Prairie Warbler by Val Nolan and one of the Wood Thrush by Jorge Vega and colleagues. These studies make clear that once the last offspring has reached independence, the breeding season is over, and for most migrant species, a new phase of the annual cycle has begun—one in which each individual is responsible for its own livelihood and well-being rather than fulfilling some role as a family member.

Using Val Nolan's 20-year body of work, we can compile a reasonably precise definition of the postbreeding period for Prairie Warblers at Nolan's Indiana study site (Figure 7.1).

Few (no?) other studies provide the quality of data regarding timing

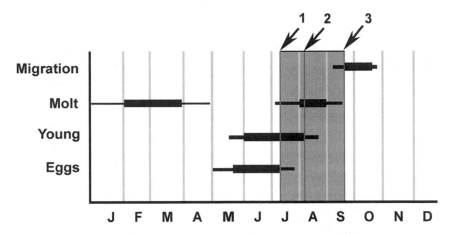

FIG. 7.1. Postbreeding period (*gray shading*) for the Prairie Warbler in southern Indiana. 1 = Mean date for independence of first brood (immatures begin postbreeding period); 2 = mean date for independence of second brood (adults begin postbreeding period); 3 = mean date for departure on fall migration by adults (end of postbreeding period).
DATA FROM NOLAN 1978

of the postbreeding study comparable to Nolan's work. Nevertheless, the hundreds of accounts in Poole's *The Birds of North America* (now included in Cornell Laboratory of Ornithology's *Birds of the World*) provide good information on when reproductive activities begin and when young reach independence. In addition, there are many good data sources documenting peak appearance of migrant passerines as transients on actual migration, such as TV tower kills, banding station activities, long-term regional data sets, and radar studies. For most migrants, these data document that the peak of migration for birds en route occurs long after completion of breeding activities. A summary combining these data for selected migrant species is provided in Table 7.1.

Assuming that the postbreeding period begins when breeding activities cease for most individuals and ends a few days before the peak of fall migration is observed, a rough calculation of mean postbreeding duration for many migrants can be made, and such a calculation is given

TABLE 7.1. Comparison of peak date for completion of breeding (based on species accounts in Poole 1992–2015) and date for midpoint of ten-day period of peak fall migration based on twenty-five years of TV tower kill data from north Florida (Crawford 1981) for selected species of North American migrants. Julian dates are given in parentheses.

SPECIES	PEAK COMPLETION OF BREEDING	PEAK FALL MIGRATION— NORTH FLORIDA	DIFFERENCE BETWEEN COMPLETION OF BREEDING AND PEAK OF FALL MIGRATION IN DAYS
American Redstart	7 Jul (188)	6 Oct (279)	91
Bay-breasted Warbler	1 Aug (213)	16 Oct (289)	76
Blackburnian Warbler	16 Jul (197)	16 Sep (259)	62
Blue-headed Vireo	31 Aug (243)	6 Nov (310)	67
Bobolink	27 Jun (178)	26 Sep (269)	91
Chimney Swift	1 Aug (213)	16 Oct (289)	76
Eastern Towhee	28 Jul (209)	26 Oct (299)	90
Gray-cheeked Thrush	8 Aug (220)	6 Oct (279)	59
House Wren	11 Aug (223)	6 Oct (279)	56
Indigo Bunting	21 Aug (233)	6 Oct (279)	46
Magnolia Warbler	3 Aug (215)	16 Oct (289)	74
Marsh Wren	15 Aug (227)	6 Oct (279)	52
Gray Catbird	3 Aug (215)	16 Oct (289)	74
Prairie Warbler	24 Jul (205)	6 Oct (279)	74
Scarlet Tanager	26 Jul (207)	6 Oct (279)	72
Swainson's Thrush	4 Aug (216)	6 Oct (279)	63
Tennessee Warbler	11 Aug (223)	6 Oct (279)	56
Veery	5 Jul (186)	16 Sep (259)	73
Wood Thrush	26 Jul (207)	6 Oct (279)	72
Yellow-billed Cuckoo	28 Aug (240)	16 Oct (289)	49
Yellow-throated Vireo	31 Aug (243)	6 Oct (279)	36

in the table, using breeding cessation data from species accounts in *The Birds of North America* and basing fall migration peak on TV tower kill data from north Florida. Data from Nolan's Prairie Warbler are included to show basic similarity in timing of the postbreeding period for this species, for which the details are well known, with several others whose details are not well known. For the twenty-one migrant species shown in the table, representing three different orders and ten families, the average length of time from completion of breeding until peak of southbound fall migration is sixty-seven days, which is probably a pretty fair estimate of duration of the postbreeding period for most songbird migrants in North America.

As Figure 7.1 shows, timing of initiation and duration of the postbreeding period will differ for members of the different age groups, and thus behaviors for members of these different major classes might be expected to show differences on average as well. We are still very early in study of this aspect of the postbreeding period, and the few intensive studies performed on migrants to date can only provide hints. Nevertheless, let us at least begin the process, starting with the adult male.

As I have said, the male arrives first on the breeding ground for most migrant species. He establishes a territory and attracts a mate, and then the pair rears the young. Often, when the young fledge, the pair divides the brood with each parent taking charge of one or two of the chicks (to feed and train) for the next three weeks or so (for songbirds), until the young reach independence. If they are raising a single brood, the male's breeding season is complete at this point, which is often by early or mid-July. About a third of the 330 or so species of North American long-distance migrants breed in forest, but by July, the insect fauna abundance in forests is dropping, whereas food resources (fruits as well as insects) in second growth and old field habitats is increasing. As a result, many individuals of forest-breeding species, adults as well as young, move out of the forest and into these habitats where they undergo the prebasic molt into basic (nonbreeding) plumage.

There is a difference between the movements and behavior of adult males and females that is indicated in the detailed accounts provided by Nolan for the Prairie Warbler and by Vega for the Wood Thrush. Adult males tend to stay near their breeding territory or, if they leave the area during molt, to return to it a week or two prior to migration, whereas females show far less tendency to remain in the breeding territory vicinity. I suggest that the reason for the difference is that males attempt to reestablish or maintain their claim on their territory preparatory for the next breeding season. At this time, one occasionally hears snatches of song in the few days remaining before the bird departs on migration, presumably to alert other males, both adult and juvenile, to the fact that this territory is taken. The data supporting this inference are fairly good for adult male Prairie Warblers and Wood Thrushes but weak for juvenile males. They, of course, have no territory of their own to return to, but I believe that prospecting for a possible territory for the coming year is a part of their postbreeding agenda, in addition to completion of the prebasic molt, of course. Nolan's Prairie Warbler accounts provide information indicative of this motivation for a single juvenile male, which he found to have established and defended a territory just prior to departure on migration.

The adult female's breeding season ends normally when the last chick under her care becomes independent. Interestingly though, Vega found more than one bird that moved to find a new mate on a different territory, where she attempted to raise a second brood while her original mate went into postbreeding mode. However, most females, adult or juvenile, simply seem to ignore territory boundaries after the breeding season concludes and move to whatever site or habitat provides the best resources to support them over the next month or two until departure on migration. Like males, although to a far lesser extent, females likely scout the region to familiarize themselves with what is available during this period, both in terms of current resources for molt and preparation for migration, but perhaps also for potential mates and breeding sites for the coming year.

The divorce between male and female roles and motivation during the postbreeding period is illustrated strikingly in migratory ducks. As discussed in Chapter 6, in several of these species, the breeding season begins on the wintering ground, where females select a mate and pair-bond for the coming year. Accompanied by her new partner, the female flies north in spring to return to her breeding pond of the previous year (or one in the vicinity of where she was raised for a second-year bird). On arrival, the pair establish their territory, select a nesting site, and copulate. Once the female has laid her eggs and begun to incubate them, the male's role in the breeding season is often over, and, for him, the postbreeding season begins. Thus, by early June, many males are free to move to the best sites or regions for molting, which may be tens or even hundreds of miles from the breeding pond. By contrast, the female remains on the breeding pond at least until the young reach independence and often through the flightless period of molt. She and the juveniles, not necessarily together, then may move to larger bodies of water to join large flocks of other ducks until migration.

The remarkable postbreeding differences between adult males, females, and juveniles in waterfowl, and probably most other migrant species in terms of movements and timing, simply serve to emphasize a critical point. Each individual is responsible for his or her own fitness. During the breeding period they share that responsibility with another individual to a greater or lesser extent, but once that period is over, they are again on their own. This information provides us with the answer to the question of why members of the different age and sex groups behave differently during the postbreeding period, which is that *postbreeding differences in movements and habitat use occur between the groups because the needs and capabilities of the individuals in these groups differ*. This situation is obvious during migration and the wintering period but, with the exception of waterfowl and some other groups, much less so during the postbreeding period.

It may not be obvious, but the finding that there are differences in terms of habitat use for the different groups also demonstrates that the *characteristics of the habitat, such as food availability, change as the summer season progresses*. Foods that were superabundant early in the season (such as caterpillars in forests) become much less so as the weeks pass, while other types of foods (such as soil invertebrates and fruits) may become more abundant in open areas or wetland sites. These changes confront the members of the various groups of any given forest-breeding species with choices. Adult males need to maintain their territories so that they have them next year, and so they attempt to stay on or near the territory site, even as the ability to find food on it becomes more problematic. Females and juveniles do not face this dilemma (although juvenile males may confront it to some extent), so they simply move to whatever habitat offers the most superabundant foods. Safety, however, is also a concern, which is why juveniles, and perhaps adults as well, often join in single- or mixed-species flocks at this time.

CHAPTER 8

POPULATION
Biology

Are not five sparrows sold for two farthings,
and not one of them is forgotten before God.

—Holy Bible, Luke 12:6 (King James Version)

Competition

This town ain't big enough for the both of us.

—*The Western Code* (1932)

A large part of our awakening to the migrant's place in its wintering environs derives from observation of the intense competition for food that occurs among migrants of the same species during the winter months. It is for this reason that study of the various behavioral techniques used to compete for food is so important. As I have said, a key assumption of the Northern Home Theory is that food is *not limiting* for migrants during the winter months. It is assumed that they can satisfy their needs simply by moving from one temporary resource bloom to another; hence, there is no need to compete either with residents or other migrants, and no need for a niche. Intraspecific competition is the driving force behind migration. Young birds, with their lack of experience, are the presumed losers in competition for food, mates, and breeding territories, and they are the likely source of the exploratory dispersal movements that lead to migration. That is why investigations of the details of competition are so important.

In order to explain how competition works in migratory birds, I will have to provide some basic theory. The equation for population growth states that change in the population (dN) over time (dt) equals the birth or natality rate (bN) minus the death or mortality rate (dN):

$$dN/dt = bN - dN.$$

Birth rate and death rate from the equation can be combined into a single term, r, as shown in the following equation:

$$dN/dt = rN.$$

Theoretically, if for any given population there is little apparent interaction between the birth rate (b), death rate (d), and population size (N), the population continues to grow regardless of the number of individuals in the population until the resources on which it depends are entirely exhausted, at which point all individuals in the population die. Bacteria on a food medium in a Petri dish approximate this situation; their population grows until the food is gone and the colony dies. This type of population growth is referred to as "density independent." Although no population is truly independent of its size, as population size is what birth and death rates act upon, in density-independent populations there is little interindividual effect on members based on their numbers alone. All individuals continue to have equal access to critical resources, regardless of their numbers, until the resources are gone. These kinds of density-independent populations are controlled mostly by random environmental factors affecting the birth rate and death rate—that is, the r term in the above equation, which is also referred to as the "intrinsic rate of natural increase" or "intrinsic growth rate."

Species whose populations are controlled principally by r are known in population biology terminology as "r-selected." This term means that the focus of their reproductive life history is on production of as many offspring as possible in the shortest amount of time.

For many kinds of organisms, like birds, the individuals in their populations differ in their ability to compete for resources, which means that some individuals can sequester resources better than others. Populations of these organisms are referred to as "density dependent," which means that individuals compete for resources when they are limited by some aspect of the environment. Pierre François Verhulst, in his 1845 paper on the topic, was the first to incorporate a limiting factor into population growth equations. His equation, often referred to as the logistic growth equation, includes a term, K, to represent the limiting factor imposed on the population by the environment.

$$dN/dt = rN(K - N)/K.$$

This equation states that a population will grow exponentially, as long as birth rate exceeds death rate, until it reaches a saturation point. This saturation point will be determined by (1) competition among individuals in a given environment or habitat, and (2) the total amount of a critical resource in that environment. Populations under density-dependent control do not die out as resources become limiting. Rather, a portion of the population is denied access to the critical resource (e.g., food, nest sites, or mates) through *competition*. This competition allows the remainder of the population to function normally, and the population achieves a steady state where birth plus immigration rate equals death rate plus emigration. The combination of total amount of a critical resource and density-dependent intraspecific competition for that resource produces a population limit called *"carrying capacity"* that is characteristic for any given population in a given environment. Theoretical population growth in populations subject to density-dependent population limitation by carrying capacity of the environment produces a sigmoid curve, as shown in Figure 8.1.

In the real world, there is a broad continuum both among and within taxonomic groups of species with regard to the degree to which life history is focused on maximizing reproductive rate (r-selected) versus those whose life history is focused on individual survival (K-selected). Normally, r-selected species occupy unstable environments so the life history emphasis is on producing the largest number of individuals in the shortest period of time, whereas K-selected species occupy more stable environments where intraspecific competition (density dependence) is important. How-

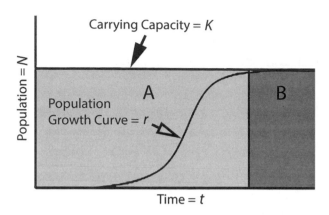

FIG. 8.1. Population growth (*r*) subject to a limiting factor (*K*) over time (*t*) beginning at a population size (*N*) of zero. Shaded area A shows a period of density independence for the population, whereas B shows a period of density dependence when population growth is controlled by competition and carrying capacity.

ever, whereas bacteria (extremely *r*-selected) and elephants (extremely *K*-selected) might be considered to be at opposite ends of the spectrum, when all species are considered, there is a broad continuum both within and between major groups of organisms. For instance, one could find examples among migratory bird species in which some appear to be more *r*-selected (such as short-lived songbirds that produce two or more clutches of three or four young per season) as opposed to long-lived species (such as the Whooping Crane, which usually produces one clutch of two eggs per season).

Population limitation for all organisms falls into one of these two categories: *density independent* or *density dependent*. For extremely *r*-selected species, control appears to be mostly or entirely density independent. However, for *K*-selected species, control can be either density independent or density dependent, depending on the status of the population size and amount of available critical habitat at any given moment in time. If more critical habitat is available than can be used by the existing population members, then the population is governed largely by density-independent factors affecting birth and death rate. Only when population size exceeds

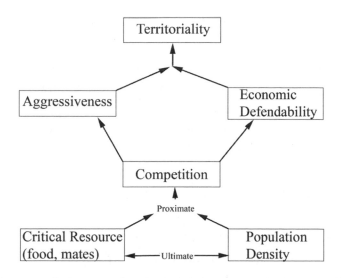

FIG. 8.2. A general theory for diversity of competitive behaviors under different circumstances. BASED ON BROWN 1964

amount of available critical habitat does carrying capacity have an effect on *r* through the medium of intensified intraspecific competition.

"Territory" is, of course, a behavioral device used by individuals to deny resources to others. In other words, it is a critical manifestation of density dependence. During the first half of the twentieth century, the term "territory," when used in the context of bird biology, was generally assumed to be synonymous with the term "breeding territory," and most discussions of the phenomenon were centered on the reproductive aspect. Jerram Brown, in his influential 1964 paper on the topic, changed that understanding by framing the argument in a different way, as shown in Figure 8.2.

Brown explains this depiction as follows: aggressive behavior is used to maximize fitness (survival and reproduction); its use should be favored when it is necessary to obtain a resource important toward the goal of maximizing fitness, such as food or mates. This situation should only occur when population density is high enough to interfere with the individual's ability to procure the necessary resource, thus producing competition. However, two additional factors affect the nature of this competition—

density of the resource and its distribution in space and time. If density of the resource is low, then it may not be worthwhile attempting to sequester it. The same may be true if the density of the resource is high, or available only for a short period of time. Energy and time spent in defense may be wasted since exclusion of competitors is not feasible. In other words, territoriality is a behavioral mechanism employed to guarantee to an individual a critical resource that is economically defendable.

Although focused on the breeding territory (the lower left box in the original figure stated "Requisites for Reproduction" rather than "Critical Resource"), Brown's attempt to explain territoriality in terms of a cost-benefit analysis, where the currency is fitness, is extremely valuable in terms of understanding the behavior in nonbreeding birds as well.

Ironically, Brown's model is much less helpful in explaining breeding territory, where the resources at issue and their relative value in the fitness equation are often less clear (mates? nest sites? food?). So if we change the label of the lower left box in Brown's figure to "Critical Resource" and recognize that the chief value for maximizing fitness in winter is survival and that food is the main contributor for achieving that end, then we can understand why the discovery and documentation of territoriality among some species of transient migrants in Texas and many wintering migrants in Veracruz rain forest was such a big deal (in my mind).

Building on Brown's theories of territory, Steve Fretwell proposed that if the amount of a given critical habitat was limited—that is if more individuals were present than the habitat could support—and if pieces of it were defended by each colonist in the form of a territory, then the habitat would become entirely occupied by territorial individuals, and all subsequent attempts at entry would be rebuffed, forcing late comers to occupy suboptimal habitat or to continue searching for suitable habitat elsewhere (Figure 8.3). He titled this model an "ideal despotic distribution." Its use was primarily to explain breeding habitat occupancy, but it seems to fit use of territoriality during the nonbreeding period as well, with the important distinction being that failure to obtain a nonbreeding territory results in lowered probability of survival rather than decreased likelihood of breeding.

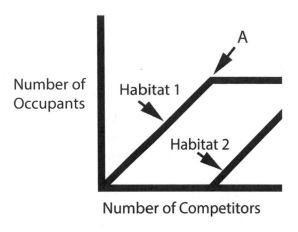

Number of Occupants

Habitat 1

A

Habitat 2

Number of Competitors

FIG. 8.3. Predicted changes in density in two habitats of differing quality with increasing population resulting from immigration using Fretwell's "ideal despotic distribution" model.

"What happens when there isn't any competition for a density-dependent organism?" That's a fair question. The answer is the same thing that happens for density-independent organisms, as shown in the graph of the sigmoid curve above (Figure 8.1). There's no dispersal, so the population grows exponentially until it reaches carrying capacity. Also, of course, there is no migration. If food suddenly becomes available year-round for a migratory species on its breeding ground, they stop dispersing *and* they stop migrating. They become year-round residents. As described below, this situation has occurred in Texas for urban populations of the White-winged Dove. Food is now available year-round for these birds in places like San Antonio with the result that their once-migratory populations are now permanent residents, and expanding exponentially. The same thing has happened recently for Dark-eyed Juncos in San Diego, California, where migratory populations have become permanent residents. These occurrences emphasize the flexibility of the migratory habit. Clearly no genetic change was required for populations to shift from migratory to resident—only environmental change in terms of food availability.

Dispersal

> A sower went out to sow his seed: and as he sowed, some fell by
> the way side; and it was trodden down, and the fowls of the air
> devoured it. And some fell upon a rock; it withered away, because
> it lacked moisture. And some fell among thorns; and the thorns
> sprang up with it, and choked it. And other fell on good ground
> and sprang up, and bare fruit an hundredfold.
>
> —Holy Bible, Luke 8:5–8 (King James Version)

The theoretical information provided above explains how competition often works for density-dependent organisms like birds. When food or other critical resources are superabundant, there is no competition. However, when resources are limited, competition ensues, and the losers, often young birds, have to leave. This departure is called "dispersal," and, according to the theory presented in this book, it is dispersal that is the engine of migration. One reason few believe this axiom is that it is almost entirely invisible. Even when we see it taking place in the form of strange sightings of members of species in unlikely places, like flamingos in Moscow, it is dismissed as an anomaly. But it isn't. Dispersal, the ability to move and reproduce away from the place in which you were born, is a characteristic of every form of life on earth. You can do it yourself (active dispersal) by flying, swimming, walking, or slithering, or you can depend on other species (predators, pollinators, etc.) or the environment (wind, water) to do it for you (passive dispersal), but without it, you're finished as a species because things are always changing, and you (or your offspring) must move and compete to survive.

Dispersal has no role in either the Northern Home or Migration Syndrome theories for the origin of migration: both theories are entirely dependent upon genetic change produced through thousands of years of natural selection on individuals in populations of temperate and boreal zone birds subjected to decreasing environmental suitability during northern winters, favoring (in a fitness sense) those with the capability of moving over those that don't.

We have had an extraordinary demonstration of the power of dispersal in recent years, but few seem to recognize it for what it is.

Regardless of what one thinks about human effects on climate, what is by no means settled is how such change could affect future climate and life on earth in general, and migratory birds in particular. For many years, I took a principally offhand intellectual interest in the research and public policy discussions regarding climate change. My sister Francesca Miller asked me to give a couple of lectures on the topic to her Stanford Washington study group in the mid-1990s, but I had no research that was relevant to the issue and no published work on the subject. In the summer of 2005, the Caesar Kleberg Wildlife Research Institute hosted a symposium celebrating their first twenty-five years; I was invited as a participant, as I had worked for several years at the Institute in the '80s. So Bonnie and I traveled back to South Texas for our first visit in a couple of decades.

It was a great meeting, Lots of time to reminisce with old friends, among whom was Dr. Lynn Drawe, director of the Welder Wildlife Foundation where Bonnie and I and our two older children had lived while I was doing my thesis research in the early 1970s. Lynn and his wife, Kay, invited us to come back to Welder whenever we could make it, and I was not long in figuring out a way to do just that, which was to do some work on a book on Texas wildlife that I was planning to write. South Texas in early spring is often quite lovely. The yuccas are in bloom, Cassin's Sparrows are singing (almost as beautiful a song as Bachman's Sparrows); temperatures are often in the 70s and 80s. So we immediately made plans to visit in February of 2006.

Shortly after we had arrived, I went out for a run along a ranch road through the mesquite chaparral. As I was shuffling along the shore of one of the oxbow lakes on the refuge, I thought I heard a Great Kiskadee. After completing my jog, I drove out to see if I could confirm the bird's presence. Sure enough, there it was, screeching away. Now, so far as I was aware, the northernmost confirmed record for kiskadee breeding was Kingsville, which is sixty miles to the south of Welder. Returning to the admin building, Bonnie and I went to the coffee room (where most discussions of importance take place at Welder). There we found a couple of wildlife biologists who worked for the Foundation. I breathlessly asked them if they knew they had kiskadees on the refuge. They did not seem very excited and said they did know. When I pointed out it wasn't on their checklist, they said, yeah, we need to update that. I asked how long kiskadees had been around. They said that they conducted breeding bird and Christmas counts every year, so they could look it up, but they figured it was at least five years. Wow. Very interesting. Later that day, I went out to see if I could find a nest. While tromping around, I heard two more unexpected species: Green Jays and White-tipped Doves. Back at coffee, I presented my information. Again, my interlocutors were not surprised. Yes, they knew about the Green Jay and the White-tipped Dove. Like the kiskadee, they had been showing up on bird counts for the last few years. I asked whether they had any ideas about what was going on. Not really, they answered. A lot of neighboring ranch land had been converted from open range to sorghum and cotton, maybe that changed something. I didn't think that seemed like a realistic explanation. Nearly all of Nueces County (the neighboring county to the south of Welder, where Corpus Christi is located) had been converted from ranchland to cotton and sorghum fields forty years earlier, and that did not seem to have an effect on the known range of any species, let alone these three. They asked what I thought was going on. I said, if it were real, such a large, recent northern shift in the ranges of three species, each of whose ecology was quite different from the others, might be related somehow to global warming. "Global Warming!" they both exclaimed. "You don't actually believe that stuff, do you? Pretty much a liberal [i.e., 'Yankee'] hoax, don't you think?" I said that I had no first-hand experience studying the phenomenon, but from what I had read, a heck of a lot of the world's climatologists did not seem to think it was a con, including some working in Texas, like Jim Norwine at Texas A&M University, Kingsville, and Camille Parmesan from the University of Texas (who shared the Nobel Prize in 2007 for her work on the issue). My friends were unimpressed.

But I was intrigued to say the least. I decided that I wanted to dig further into the question of range change, which I define as, "the appearance of a species as a regular part of the breeding fauna of a region from which it was not known to breed based on historical documentation, or its disappearance from a region where it was known to breed."

There are two key parts to understanding such change: good historical documentation of what the former range was and current data documenting the new range. Consequently, the beginning for my project was to check my observations against the historical record. Fortunately, in Texas there are excellent accounts for most bird species dating back to the mid-1800s, thanks largely to the Mexican War (1846–1848), which brought some excellent field biologists to the region, mostly as army surgeons. In particular, the bird communities of both the Kingsville and Welder areas have been intensively studied for at least the past seventy years. As described earlier in this book, Bonnie and I captured more than ten thousand birds at Welder from March through May and August through October for two years during my thesis work back in the 1970s. In addition, I did bird research at Texas A&I University (now Texas A&M University) in Kingsville for eight years, from 1981 to 1989.

Therefore, I knew, based on my own work, that Green Jays and kiskadees *were breeding* in the Kingsville area in the early '80s and that they *were not even present* at Welder. In addition, my good friend, colleague, and coauthor of two books on Texas bird distribution, Gene Blacklock, worked for twenty years as the curator at Welder, and he had no records for Green Jays, kiskadees, or White-tipped Doves from that same time period of the '70s and '80s, despite frequent surveys, and these birds are hard to miss, being quite vocal as breeders. Thus, apparently between the time that I left Texas in 1989 and the time I returned in 2006, three species of birds had extended their breeding range sixty miles to the north.

That intriguing piece of information stimulated me to conduct a detailed literature search to determine whether there was information on range change for other species. As I have mentioned, Texas is fortunate in having excellent historical documentation of bird distribution. I used two main sources for my investigation: *The Bird Life of Texas* by H. C. Ober-holser (brought up to date by senior editor, E. B. Kincaid), which provided detailed range information for every Texas bird up to its date of publication in 1974, and *The Texas Ornithological Society (TOS) Handbook of Texas Birds* (coauthored by Mark Lockwood and Brush Freeman), which provided range information current up to its date of publication in 2004. To summarize my findings from analysis of distributional records in these works, I found that *at least eighty species of tropical, subtropical, and warm desert bird species appeared to be in the process of shifting their breeding ranges hundreds of miles north and east into completely new ecological regions.*

When I first began conducting these studies, I contacted a friend of mine who had devoted his career to climatology of the Texas subtropics, the aforementioned Dr. Jim Norwine at Texas A&M University in Kingsville. I asked him if he might be interested in providing some input on climate change in this region to go along with my observations on breeding bird distribution changes. He said, funny I should ask, because he was in the final stages of editing a book on the topic. He invited me to submit a contribution on my birds. And he sent me his introductory chapter for the volume wherein he and his coeditor, Kuruvilla John, state that the *climatic changes currently underway in South Texas (decreasing mean precipitation and increasing temperature) are likely to have an effect over the next century comparable to moving the region 100 miles to the southwest*, essentially converting it from a humid, subtropical region to a warm desert. When I saw this statement, it seemed to me that I had found evidence indicative of a possible relationship between the climate change predicted by Norwine and John and changes in species distribution in the form of northeastward shift in breeding range of eighty tropical, subtropical, and warm desert bird species. But, were these range changes real? Sightings alone do not provide sufficient data for documentation, and all of the information on range change summarized in the aforementioned TOS Handbook came from sightings by amateurs rather than specimens (skins, skeletons, and other remains of actual birds in museum collections), which is what distribution was based on in *The Bird Life of Texas*.

FIG. 8.4. A member of the Rock City Bigfoot population, caught on film near Olean, New York.

As an illustration as to why sightings alone can be a problem, consider that the nonhuman primate with the broadest distribution in the world based on sightings is, drum-roll, maestro . . . *Bigfoot* (Figure 8.4). In other words, it takes more than sightings (visions? apparitions? recovered memories?) to properly document such a monumental occurrence.

Welder's location at the northern end of the subtropics seemed to me to be the perfect place for conducting an investigation into avian range change, which I did for breeding seasons from 2007 until 2010. The hypothesis for this research was as follows:

Individuals of subtropical bird species observed north of their historic breeding range at the Welder Wildlife Refuge represent new populations forming a northward extension of the breeding range.

To test this hypothesis, I used the following methods:

1. Record exact global positioning system (GPS) locality for singing males or other observed evidence of pair formation for all bird species not found historically as regular breeders at Welder according to Gene Blacklock's 1984 *Checklist of Birds of the Welder Wildlife Refuge.*
2. Use playback of recorded calls to test for territorial response (approach within thirty feet by one or both members of a breeding pair).
3. Capture and band with numbered aluminum bands (from the National Bird Banding Laboratory) and plastic color bands to determine seasonal and annual persistence by resighting and recapture.
4. Establish a feeding station monitored by observers via the internet who record presence of banded birds by photo.
5. Collect nests, eggs, or nestlings of target species (those whose breeding ranges terminated south of Welder as documented in *The Bird Life of Texas*).

Results of this work showed the following:

1. Four subtropical resident species had made a northward shift in breeding range of sixty to one hundred miles in a period of thirty years based on observation of nests, eggs, or fledglings: the White-tipped Dove, Green Kingfisher, Great Kiskadee, and Green Jay.
2. In addition, seven species of tropical-subtropical migrants had shown northeastward breeding-range shifts: the Groove-billed Ani, Black-chinned Hummingbird, Buff-bellied Hummingbird, Couch's Kingbird, Cave Swallow, Audubon's Oriole, and Bronzed Cowbird.

These data also place the anecdotal information in the TOS Handbook in a different light. It is now apparent that where the new range suggested by this publication is based on numerous sightings over several years, the range has a high probability of correctly representing an actual range expansion. If this conclusion is correct, then a careful reexamination of the range information in the TOS Handbook and a comparison with previously published information are obviously warranted. The shift in range for eighty species, although based on sight records, take on the attributes of a new hypothesis with powerful support of its accuracy based on the data collected on nine of those species at Welder.

All right. So eighty species of subtropical birds have shifted their breeding ranges as much as one hundred miles to the north in less than thirty years. How is this possible? Dispersal, obviously. It's always there. It never stops. So when areas that previously were uninhabitable become suitable, for whatever reason, there are always dispersers to find it. What happens to those that don't find it? Easy. They die "on the road," or they try to go back home, if they can, to see if things might have improved for them since they left. Either way, short-term range change is solid proof that dispersal is an important part of any population that produces more individuals than resources can support, which is true for most density-dependent organisms living in relatively stable environments.

The study of range change in the Texas subtropics provides some additional insight, I believe, into how dispersal becomes migration. As noted above, four species (Green Jay, White-tipped Dove, Great Kiskadee, and Green Kingfisher) expanded their range northwards, with populations occupying the new range as full-time residents. This behavior demonstrates that food that had not been available to them year-round previously now is. However, seven species expanded their ranges northward only as breeding populations. Once the reproductive season is complete, they head south. Why don't they stay? Presumably because the foods that they need are *not* available throughout the winter period. This finding illustrates that successful dispersal equals range expansion if year-round residency is possible, but it is migration when the dispersers must leave the new-found breeding site if seasonal change causes critical foods to disappear. When this happens, they simply return to where they came from.

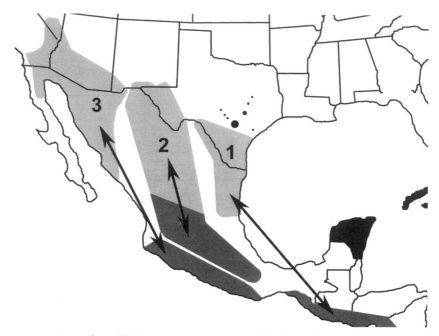

FIG. 8.5. Breeding (*light gray*), wintering (*dark gray*), and resident (*black*) populations of the White-winged Dove. 1 = the South Texas breeding population; 2 = the West Texas breeding population; 3 = the Arizona and northwestern Mexico breeding population. Arrows denote migration to principal wintering regions for each population. Note the black dots signifying resident populations in major Texas cities, the largest dot representing the huge resident population of more than a million birds in San Antonio.

The White-winged Dove illustrates how change in seasonal food availability can change migration behavior in very short order (no genetic change required). Whitewings are a common breeding species of southwestern thorn forest that winter in drier portions of southern Mexico and Central America (Figure 8.5). In the late 1840s, Texas whitewing populations were found no farther north than the Rio Grande Valley, but by the mid-1970s, they had expanded their breeding range more than one hundred miles to the north. Why? Presumably because of dispersal. Food availability during the breeding season changed in that region, allowing dispersing birds to establish breeding populations in vast new parts of South Texas. Whether this change was a result of climate change or agricultural practices or both doesn't matter. Food availability changed and the birds responded. Note that the new populations were migratory, just like their parent populations, again presumably because foods that were available during the breeding season were not available in the winter season. Since the 1970s, whitewings have extended their populations another three hundred miles northward (and still moving). However, these populations are not migratory. They are resident, and they are found only in cities—not out in the countryside. Why? Again, dispersal is the answer. But why aren't they migrants? Because they don't have to be. The foods that they depend on, mostly seeds, are available year-round at backyard feeders, grain elevators, and so forth.

The findings reported on rapid range change for South Texas birds firmly document the extraordinary power of dispersal. This observation comes as no surprise to many field biologists, but it has profound implications for theoreticians. Most of the equations devised to explain how the process of speciation works assume that the range of a species is sacrosanct. If some individuals should manage to wander outside the range, it is assumed that they don't get very far, and even if they do, they do not interbreed successfully with members of populations that are closely related. For instance, this assumption is fundamental to Ernst Mayr's definition of the "biological species," namely "groups of interbreeding populations reproductively isolated from other such groups." For theorists, geographic isolation (that is, ranges separated by large distances or habitat or topographic barriers) is necessary to explain how genetic differences between closely related populations occur. What the Texas data demonstrate is that distance is not a barrier for dispersing individuals if food is available along the route. Which means that some other factors besides geographic isolation have to be figured out in order to explain how new species originate. This need has been obvious for some time to field biologists in Texas, where a large number of pairs of closely related species exist with little more than a river separating their ranges.

Population Limitation

> No room in the inn.
>
> —Paraphrase from Holy Bible, Luke 2:7 (King James Version)

In mid-April 1966, a symposium was held at the Smithsonian Institution to which the leading experts in neotropical ecology and ornithology had been invited. All the big names were there: Ira Gabrielson, Marston Bates, S. Dillon Ripley, Eugene Eisenmann, Alexander Wetmore, and Leslie Holdridge, among many others. William Vogt was the force behind this gathering. Observations made during his travels throughout the hemisphere as scientist and diplomat had convinced him of the Malthusian inevitability that *untrammeled growth in human population and its attendant devastation of natural environments posed an existential threat to North American migratory birds that wintered in the neotropics.*

The volume resulting from this convocation, *The Avifauna of Northern Latin America* edited by Buechner and Buechner, is instructive. Despite the ostensible topic, none of the participants other than Vogt himself expressed the slightest concern regarding migrant vulnerabilities. In fact, as the title suggests, most paid little attention to the proposition under consideration in their talks, feeling free to discuss whatever their current research interest happened to be, so long as it pertained in some way to birds. Allan Phillips and Roger Tory Peterson (a native of Jamestown, New York, like me, by the way) were among the few to actually address the main issue in their comments, which can be summarized as follows: "No need to worry about migrants. They mostly spend the winter in scrubby second growth, or even residential areas, so tropical habitat loss should have little or no effect on their populations."

This belief was derived not only from what little information there was on migrant behavior in the tropics but was also soundly based on the extant paradigm for migrant evolution, which held that migrants were temperate zone species whose biology and behavior were shaped almost entirely by their temperate (breeding) environment. Back in 1966, at the time of the Smithsonian convocation, nearly all ornithologists considered migratory birds to be bona fide members of density-dependent species whose populations were held at relatively stable levels by competition for *breeding territories*. Naturally, it was assumed that the main threat for any migrant species, but especially songbird migrants, was reduction in the amount of breeding habitat. And, in fact, this seems self-evident. Each year birds raised their offspring on their breeding territories and then left, returning the following spring. As long as appropriate habitats were available, they came back. If the habitat was gone, they didn't. So clearly, breeding habitat is the starting and ending point of concern regarding migrant population dynamics. As discussed in Chapter 1, Robert MacArthur's theories were designed to provide biologically realistic models for how this system worked. Migrants arrived, defended their territories, raised their young, and, when weather got bad, left. During their southern sojourn, they survived essentially as fugitive or pioneer species, continually moving from one concentration of superabundant resources to the next until the weather moderated and they could return. No worries about competition or carrying capacity because they were always on the move. If someone were so naïve or uninformed as to inquire about how they survived the rest of the year, they just didn't understand how migrant populations worked. Thus, Dr. Vogt's worry that migrants might be threatened by winter habitat loss was basically dismissed by the ornithological community on both experiential and theoretical grounds.

Our field work in Mexico in 1973 left me far less sanguine. We looked long and hard to find good forest, and when we finally located some in southern Veracruz, we discovered that members of twenty or so species of North American migrants appeared to be dependent upon it for their overwinter survival. Whatever else the findings from our first year of work in the tropics might mean, I felt that our data demonstrated the possibility that William Vogt's concerns regarding wintering habitat loss might well be correct. I didn't hesitate then to take advantage of an offer from the Welder assistant director, Eric Bolen. After visiting our Mexican study site in early 1974 with Dr. Cottam, Bolen became excited about our work and suggested that I submit a paper to the *Bulletin of the Texas Ornithological Society*, which was edited by a good friend of his, Dr. Kurt Rylander. The bulletin was a regional journal with a small readership of mostly nonprofessional bird watchers. Nonetheless, it had the potential to give me a chance to put important early findings from our work out in the intellectual marketplace, and I jumped at it. Dr. Rylander eventually accepted a revised version of my short article, and the paper came out in the December 1974 issue under the title, "Migrants and Space: The Wintering Ground as a Limiting Factor for Migrant Populations." This paper provided a brief summary of the results from our work in Texas and Mexico regarding long-term site fidelity of migrants to their tropical and subtropical wintering habitats, and it proposed that, if such habitats continued to be destroyed at the alarming rates reported by environmental agencies, populations of migrants dependent upon them would suffer.

There were at least a couple of positive results from this effort. First, it gave me my first publication. Quantity and quality of publications are the basic measures of accomplishment for any research scientist. Second, it stimulated a long appraisal from Dr. Frederick Gehlbach, a professor at Baylor. He sent a letter commenting on my paper that was highly critical of my Mexican findings, citing in particular the work of Elliot Tramer, University of Toledo biology professor (and MacArthur disciple). Dr. Tramer's paper, "Proportions of Wintering North American Birds in Disturbed and Undisturbed Dry Tropical Habitats," addressing the same topic of migrant wintering ecology as my own, and also based on work in Mexico (Yucatán), reached conclusions quite different from mine. He found that migrants were among the most common inhabitants of gardens and residential areas of Merida, but they were also common in the native dry forest that covers much of the northern part of the Yucatán Peninsula. He noted that migrants were seldom or never found in stable tropical habitats like rain forest. Dr. Gehlbach averred that certainly I must be mistaken in my findings, which were obviously contrary to Dr. Tramer's, as well as most previously published information on wintering migrants in the tropics.

Initially dismayed, I eventually realized that if my findings weren't important, a person as prominent in his field as Dr. Gehlbach would not have bothered to critique them. Also, I found that I was able to rebut his critique using data from our Mexico work. But perhaps most importantly, Gehlbach's letter underscored a fundamental puzzle of wintering migrant biology, namely how could both Tramer and I be right? How could wintering migrants of the same species behave in one way in residential areas, pastures, and second growth (small, loose flocks of wanderers) and another way in primary forest (solitary territorial individuals showing long-term site fidelity). I had an inkling back in early 1975, but it took me many years of field study, collaboration, and discussion with experts on migration from other parts of the world before I was able to develop a hypothesis that could reconcile these seemingly inconsistent sets of observations.

What I failed to recognize in my naïve publication was that the idea of migrant populations being limited by wintering-ground factors had the potential to turn the entire field of migrant population dynamics upside down (and migrant conservation as well). The whole notion of a Northern Home is based on the assumption that migrants are adapted to, and dependent upon, their breeding habitats. This proposition includes the corollary that migrant ecological and behavioral adaptations focus on survival and reproduction on the breeding ground. Once they leave the breeding ground in fall, they have left their niche (that conceptual space where they can compete successfully with all other species) and become dependent upon location and exploitation of superabundant resources for the next several months until their return to the breeding area in spring. My claim that migrant populations were dependent on wintering habitat for survival brought me directly into conflict with the vast majority of working ornithologists, most of whom were involved with field investigations of various aspects of migrant breeding biology and natural history, studies that were based on the assumption that this portion of the annual cycle was key. Furthermore, it begged the question of what the heck actually was happening on the breeding ground. If migrants originate from the wintering area, and their adaptations, their niche, derive from competing to survive during that portion of the annual cycle, then what is the importance of the breeding area? Where does it fit in terms of the migrant species' population dynamics and evolution?

My understanding of these questions was superficial to say the least back in 1974, but our LaSalle Creek Bog study conducted in Minnesota during the summer of that year actually held the crucial elements to address these questions. At the time of our studies, the breeding territory for most migrant and resident landbirds was defined as a piece of ground defended by the breeding male for the *exclusive* use of the breeding pair for the purposes of feeding, mating, nesting, and rearing of offspring (the "Type A" territory, as defined by Margaret Morse Nice). The fact that the resources of the territory are defended solely for exploitation by the owner and his family in this definition is the critical point. If true, it means that the population size of a migratory species must be determined by the number of breeding territories available, or to put it another way, amount of breeding habitat establishes the carrying capacity; nonbreeding habitats are not important in this regard.

But we found that migrant breeding territories at LaSalle Creek Bog were *not* exclusive, and we were by no means the first investigators to discover this fact. The presence of nonbreeding birds ("floaters") on the Type A breeding territories of many species of migrant songbirds has been known for over a century and a half. In fact, it was that paragon of field biologists, Charles Darwin, who was the first to my knowledge to document the phenomenon in the literature. What was not recognized was the contrast in behavior toward floaters for males defending breeding territory as opposed to males and females defending feeding territories during migration or on the wintering ground. On the breeding ground, we found that territorial males mostly ignored the presence of floater males, the only exception being if the floater vocalized or attempted to approach the territory owner's mate, in which case it would be attacked and chased. This behavior is in stark contrast to what we observed on transient and wintering territories, where any intruder, regardless of sex, normally was attacked and chased as soon as it was detected by the owner.

The presence of large numbers of floaters, and the lackadaisical attitude of territory owners toward them, indicates that food is not a limiting factor on the breeding ground and that the purpose of the territory is not primarily to secure a food supply for the sole use of the pair and their brood. There are other such indicators. Jared Verner, for instance, in a paper published in 1977, summarized a large amount of literature on migrant breeding territories documenting that for many songbird migrants breeding territory size was not a fixed entity. Instead, a clear pattern existed in which the arriving male initially establishes a territory that is very large in size preparatory to the female's return. Once the female has arrived and pairing and nesting have occurred, the defended territory shrinks to half of its earlier size or less. When the young hatch, at a time when the food needs presumably are greatest for the family, almost no time at all is spent in defense, and only a small percentage of the originally defended area is actually used for locating food to feed the young. After the young have fledged, the adults often split the brood with each taking charge of one or two fledglings, and they set off to forage wherever they find the richest resources without paying any attention whatsoever to territory boundaries or to any other individuals of the same species. This pattern of territorial defense does not sound like it is tied to food availability.

Exemplary experiments conducted by Lars von Haartman and his students on European Pied Flycatchers nesting in Finnish pine forest raise further questions regarding the role of food as a determinant of breeding population size. These birds breed across northern Eurasia and winter mostly in central Africa. Preferred nest sites are woodpecker holes, but von Haartman found that the birds would use nest boxes quite readily, and that breeding population size could be significantly increased simply by adding nest boxes. Since that work, the same discovery has been found to apply to many other hole-nesting species, most notably the Eastern Bluebird in the United States. This finding demonstrates that it is nest site availability that determines breeding population size, not food availability, at least for these species.

These observations lend further support to the supposition that provision of a stable food supply is not the main reason for the breeding territory.

If that is true, then these discoveries essentially reverse the existing model of migrant population dynamics—little evidence of breeding-ground population limitation and considerable evidence of wintering-ground population limitation, posing a considerable conundrum for migratory bird biologists clearly illustrated by heated exchanges in the literature lasting from the 1970s up to the present day.

The cause of these duels was the discovery that North American populations of several songbird species did, in fact, appear to be declining as predicted by Dr. Vogt. Two schools of thought quickly materialized in publications on the topic: (1) those who believed that the declines could be explained by breeding-ground factors, and (2) those who believed that the declines could be explained by wintering-ground factors. A third group claimed that the declines were, for the most part, statistical anomalies.

The "breeding-ground limitation" camp proposed that declines resulted either from breeding habitat loss, "habitat fragmentation," or a combination of the two. The fragmentation concept holds that when a bird's breeding habitat is changed from large blocks to small, isolated pieces, two things happen. First, limitations to populations in pieces occur based on island biogeography theory (as developed by Robert MacArthur and E. O. Wilson), namely that populations in the piece become governed by probabilities of immigration and extinction based on size of the piece and distance from a large habitat block. Secondly, they argued, the process of breaking a large block into smaller pieces involves changes in the fundamental nature of the original habitat that reduce its quality for breeding purposes.

Many studies were done on the habitat fragmentation phenomenon and its effects on birds, mostly in the United States. Some of the findings were incontrovertible: decrease in block size caused increase in nest predation (mostly by chipmunks, mice, Blue Jays, and the like) and social parasitism (by Brown-headed Cowbirds). But two findings were odd. First, *size* of the site alone seemed to have a negative effect on breeding population; that is, the smaller the site, the lower the density of breeding pairs. Second, this effect of size seemed to apply only to migrants; resident populations did not show density declines related to size of the habitat piece. Chan

Robbins and colleagues published the most famous documentation of this "area effect" in a 1989 paper in the journal *Wildlife Monographs*, where they actually presented data on the size of habitat blocks below which no breeding individuals of a given migrant species could be found for their Maryland study area.

My colleague Vicki McDonald and I published a paper in *The Auk* in 1994 wherein we addressed the seemingly intractable question of whether or not migrant populations were limited by breeding versus wintering-ground factors. Accepting the extreme difficulty of actually answering this question through measurement, we proposed that one could, instead, make a set of predictions based on the assumption of when during the annual cycle a migrant population was limited, and then one could examine available information to determine whether or not the assumption was supported by real data from the field. For example, consider the issue of optimal habitat occupancy by season. If we assume that populations of a given migrant species, say, the Wood Thrush, are limited by breeding habitat, we would expect to find that all available high-quality breeding habitat would be filled, and that individuals would be found breeding in habitats of lower quality as well, as predicted by Steve Fretwell's "ideal despotic distribution" hypothesis (see Figure 8.3). However, this is not what we find. In fact, what Chan Robbins's "area effect" would seem to indicate is that vast areas of suitable breeding habitat are unoccupied based on size alone, and not only for Wood Thrushes, but for many other migrant species that breed in forest.

Applying the same assumption to winter habitat, if optimal wintering habitat were not available, we would expect to find Wood Thrushes, or any other forest-related migrant, in a variety of suboptimal habitats, which of course, is exactly what we, and many other observers, have found. Wood Thrushes prefer to winter in lowland wet tropical forest, as demonstrated by the fact that they defend and live on territories in such forest throughout the winter and return to their territories in subsequent winters. Nevertheless, many Wood Thrushes are found in scrub, roadside tree lines, borders of agricultural fields, and similar degraded environments in winter, where they suffer mortality rates much higher than those observed in Wood Thrushes wintering in tropical forest—an outcome which we would predict if winter habitat were limiting.

We presented a total of fourteen predictions of the type described above, in which actual data from breeding and wintering populations could be used to evaluate where in the life cycle population limitation might occur. For all fourteen, the wintering ground appeared as the most likely period, at least for the forest-related migrants that were the main focus of our own work.

This paper was the first skirmish of what has turned out to be a long and bitter war. Regardless of the logic of migrant wintering behavior, it is very difficult to measure migrant populations in ways that can provide unequivocal documentation of whether or not they are increasing or decreasing. Not only that, if you are able to document significant decrease, it is extremely difficult to determine where in the annual cycle the decrease is happening.

The first problem, that of accurate measurement of population trends, was addressed by an extraordinary government effort in basic science, originated by Chan Robbins and sponsored and staffed initially by the US Fish and Wildlife Service's Bird Banding Laboratory at Patuxent, Maryland (now under the aegis of the US Geological Survey). Robbins designed and organized a nationwide breeding bird survey in 1966, which has been conducted annually since then. The survey is conducted by thousands of volunteers across the country. The data are necessarily quite crude and cannot be used to calculate population size. However, they can be used to determine trends, that is, whether or not a population is increasing or decreasing. By the 1990s, these data showed that many populations of migratory birds that wintered in neotropical forests were decreasing.

But what was causing the decrease? I argued in my papers that the decreases obviously were caused by loss of wintering habitat, which was eighty per cent or more for Middle American forests. Others, by far the vast majority of students of the problem, argued that breeding season factors were the cause for most migrant declines. In fact, the suggestion that populations of migrants might be controlled by factors outside the breeding ground was considered by many to be heretical and contrary to predominant theory of how migrant population ecology works.

Ironically, because of his ground-breaking theories on importance of the nonbreeding season to migrant populations, it was Steve Fretwell who most famously described and depicted how territoriality might function to control population size during the breeding period in his ground-breaking book *Populations in a Seasonal Environment*. His model of ideal despotic distribution proposed that for a given migrant species appropriate breeding habitat could be considered to amount to a specific area. This space would fill with pairs, and the males of each pair would defend a territory of a specific mean size. Once the number of territories defended equaled the total size of the breeding habitat, no more pairs would be allowed to settle due to competition from existing occupants (this is where the "despotic" comes in).

Think of the total amount of breeding habitat as a checkerboard with each square representing the average size of a single breeding territory. The board has sixty-four squares, and once sixty-four pairs have settled, the habitat is full. No more pairs will be permitted; a classic example of density-dependent control of population size.

As we have discussed earlier, and as many, many other researchers have found, occupation of breeding habitat for migrants actually seems to work something like what Fretwell described, although not quite in the ideal way shown in the model. For instance, in the model, once the sixty-four pairs are on territory, the other birds simply disappear, which we know does not happen in real life. Actually, we know that at least three things happen depending on the age and sex of the individual: (1) many males unable to locate suitable habitat for territory establishment act as floaters, continuing to explore established territories, taking over a territory if the owner disappears; (2) other males, especially younger, second-year males, establish a territory in habitat of lower quality and attempt to attract a female with which to pair; (3) females unable to locate an unpaired male in suitable habitat settle on the territory of a paired male, accepting a polygynous pairing rather than forgoing breeding for the season. These findings demonstrate that the carrying capacity concept is not quite as clean as the model would indicate. For one thing, carrying capacity on the breeding ground bears little or no relationship to individual survival of adults. Food is not a problem for most during the breeding season. Thus, the resource referred to in breeding-period carrying capacity is the territory, or more specifically, what the territory represents, which is the possibility of raising offspring to maturity.

These are quibbles, though. From a population ecology perspective, Fretwell's models and equations demonstrate how amount of breeding habitat can serve as a density-dependent control on population size for migratory birds. Fine. That is what population ecologists have argued since the field was formulated. But what if populations are below carrying capacity? What if apparently suitable breeding habitat for a species is not occupied, as has been observed recently (over the past thirty years) in a number of long-distance migrants? The explanation normally given is that density-independent factors (such as hunting, pesticides, cats, disease, wind farms, in-transit dangers, etc.) kill more birds during the nonbreeding season than the breeding habitat can support. Our Mexican work, however, suggests another possibility. What if migrants encounter more than one carrying capacity over the course of the annual cycle? What if there is more than one checkerboard where they have to defend their own square, not in order to breed, but to survive? This possibility is what our Mexican research suggests.

As luck would have it, a sequence of events back in 2005 put me in touch with just the right person to address how this might work from a theoretical perspective. Early that summer I was contacted by the editor of the *Audubon Naturalist Society Newsletter* asking me to submit a short article on factors affecting migratory bird populations. I titled my article "Every Sparrow That Falls," alluding to St. Luke's passage quoted above (the editor changed it to "Not All Deaths Are Equal"). In it I explained that, while bird deaths were personal to many people—especially if they happened in front of their eyes, as when a neighbor's feline tortures and disembowels a newly fledged baby robin—such deaths likely have nothing to do with population size, which for most migrants is a function of available habitat, either on the breeding or wintering ground.

A reader took exception to my position, claiming in a letter to the editor that habitat was only one of several factors to be considered in determining what might affect population size for a species. In my responding letter, I explained the concept of carrying capacity, concluding that, if a population were below carrying capacity for its habitat, then my critic was correct—every death had an effect. However, using the Verhulst equation, I tried to explain that if habitat carrying capacity were the limiting factor, then the various causes of death were immaterial. All individuals excluded from the habitat would die anyway—of predation, starvation, or whatever. The cause of death didn't matter. So far so good. Standard population theory. But then I went a bit further to explain that, for migrants, there could be more than one habitat and more than one carrying capacity. I did not actually know this, nor had anyone ever suggested it so far as I was aware. But, hey, it made sense.

This response prompted a communication to me from Dr. Alan Pine, an avid birder and retired engineer (UC Berkeley PhD and Harvard postdoc with a career at MIT Lincoln Laboratory and the National Institute of Standards and Technology in high-resolution gas-phase molecular spectroscopy, tunable lasers, nonlinear optics, Raman and Brillouin scattering in solids and liquids, microwave acoustics, cryogenics, mid- and far-infrared technology, yadda yadda—you get it; a smart guy).

Dr. Pine advised me in his epistle, in the nicest way possible, that the situation with regard to multiple carrying capacities was probably a bit more complicated than I had intimated. I then called him up and asked if we could meet. We did, and I asked him if he would help me to figure out just exactly how multiple carrying capacities might work. He agreed, which initiated a long and fruitful "collaboration" (he did ninety-nine percent of the work—which is typical for me, as Bonnie and many of my students would say) culminating in our paper on White-winged Dove population ecology, published in the proceedings resulting from the Kleberg Institute twenty-fifth anniversary celebration, and the wonderful set of equations discussed in, and presented in the appendices of, *The Avian Migrant*.

What Alan found, after months of hard digging in the literature and fancy high-powered mathematical modeling, was that the situation regarding multiple carrying capacities for migratory species was, indeed, complicated. For those with the knowledge and interest, the best source, of course, is the equations themselves with their accompanying explanations, as given in the chapter on population ecology in *The Avian Migrant* along with the relevant appendices in that volume (corrections provided in Appendix 3 herein). However, to summarize from that source, "the smallest carrying capacity encountered over the course of the annual cycle has the most influence on the ultimate population size." Just like I said. Sort of.

How such models function out in the real world depends on the species and its specific mode of life during the time it spends in the various habitats encountered during the year, and it is the real power of Alan's equations that they illustrate this circumstance clearly. Let us take, for example, one of our long-distance migrants from southern Veracruz, the Hooded Warbler. Population size during the breeding period for this bird results from a combination of immigration (all adults in the population are assumed to be immigrants from their tropical wintering ground), births, emigration, and deaths. Formerly, that is, before Alan's contributions, carrying capacity (total number of birds that the habitat could support) was assumed to be determined for all species of migrants by the number of territories that the breeding habitat could hold, which, in turn was assumed to be a function of food availability. Thus, using the checkerboard model of ideal despotic distribution, all we have to do to estimate total population size during the breeding period for this migrant is simply take the total amount of breeding habitat (size of the checkerboard), divide it by the average territory size, multiply that number by two (the male and female), and add the average number of young raised to maturity, and voilà, we know the carrying capacity.

Actually, this reasoning is still used in most (all?) population modeling of migrants for conservation purposes; that is, total amount of breeding habitat is the key number used to calculate what small populations require in order to have a chance of survival. For that purpose, it makes some sense. After all, there *is* a relationship between total number of breeding territories and population size. Nevertheless, it is important to remember that at least three of the fundamental assumptions of the calculation may be incorrect: first, that food is limiting during the breeding period; second, that the territory size bears some clear relationship to the amount of food required for the family; and third, that the territory is occupied solely by the family members. For the Hooded Warbler, and most long-distance, forest-breeding songbirds, we know that these assumptions are false (as discussed in Chapter 6): food usually is *not* limiting during the breeding period, food is *not* the determinant of territory size, and nonbreeding adults often *are* found on the territories of breeding pairs.

One of the beauties of Alan's carrying capacity equations is that they take these issues into account. Each key determinant of population size is assigned a factor in the equation used to calculate carrying capacity for each habitat occupied by the population over the course of the annual cycle. Thus, for each of the habitats used for breeding purposes, an obvious factor would be food availability, another might be nest site availability, another could be predation risk or disease or pollution, or availability of mates, and so on. The value of the equations in this regard is that they serve as a way of indicating what elements of life history are critically important. Determining what those are would be the task of the field researcher.

Another brilliant aspect of the Pine equations is that they take into account the fact that carrying capacity can mean different things for different age and sex groups within the population. This observation has been well documented in a number of migrant species: older males hold the majority of breeding territories in high-quality habitat; younger males comprise a larger proportion of nonbreeding floaters discovered on breeding territories, as well as occupying territories in poorer-quality breeding habitats (that is, those in which there is lower probability of attracting a female); females of all age classes seldom occur as floaters because they can always accept a polygynous relationship by settling on an occupied territory; and so forth.

A third ingenious quality of Alan's work is that it allows for inclusion of different carrying capacities for different habitats occupied, not only during different seasons but within a season as well. In other words, we know perfectly well that, for many species, carrying capacity can be quite different in one part of the breeding range as opposed to another, and it is important to our understanding of the species' population ecology that we recognize such differences.

I have emphasized how Dr. Pine's work helped expand our understanding of breeding season population ecology for migrants. But, of course, the main focus and driver behind his studies was to model how nonbreeding-season dynamics might affect migrant populations. To do this, he developed equations similar to those used for the breeding period, where each habitat occupied—whether as a transient or as a winter resident—was assigned a set of factors, each of which had its own density-dependent contribution to controlling overall population size. Following this reasoning, let us consider the various parts of the nonbreeding period.

The first such nonbreeding habitat occupied would be that of the postbreeding season. As noted in Chapter 7, postbreeding habitats can be quite different for the different age and sex groups. Adult males tend to stay on or near their breeding territory or, if they leave for the molting period, return at some point later in the season for a week or two to the territory prior to departing southward on migration. Juvenile males also seem to spend some of the time of this postbreeding period prospecting for next year's breeding territory. Food does not appear to be a limiting factor during the postbreeding period for most migrants, although few have been intensively investigated. Therefore, the likelihood of density-dependent effects on the population seems small.

The second major habitat occupied during the nonbreeding period is the sky. Clearly, there is no carrying capacity for this habitat, and no birth rates. However, there are immigration, emigration, and death rates, and they differ for different routes through the sky and for different age and sex groups. Several investigators have assumed that this habitat is the most dangerous for migrants, resulting in the highest death rates and potentially keeping populations of migrants below the carrying capacity for any other critical environment encountered over the course of the annual cycle. Certainly it is true that many migrants die in storms or as a result of headwinds encountered over water or deserts, but what these deaths constitute as a portion of any given population is unknown. Considering return rates of sixty percent or more for migrant Hooded Warblers to their Tuxtla winter territories subsequent to two trips of a couple of thousand miles each way, including at least one across the Gulf of Mexico, one can see that migratory flight may not be quite as dangerous as we imagine. In addition, the fact that individuals of many migrant species *do* show density-dependent competition (defense of space) for food at times during migration (at stopover points) and often during the wintering period is indicative that no density-independent element holds these populations below carrying capacity.

The third major habitat, or really a succession of many different habitats, is stopover areas. Most songbird migrants in North America travel during the night (fewer predators? less air turbulence?), stopping to rest and feed during the day. As I have explained in earlier chapters concerning my work at Welder, I spent a great deal of time investigating the population dynamics for several species of migrants observed during stopover in riparian forest and mesquite thorn forest. I came to believe that there were two different classes of transients: (1) birds in more or less of a flying state (*Zugstimmung*), ready to continue their migration that night and not particularly selective with regard to habitat, which remained social during the day, and (2) those in a feeding state (*Zugdisposition*), actively trying to rebuild their fat reserves and therefore highly selective with regard to habitat and intolerant of conspecifics. This situation might have been true for a few of the migrants, such as the Northern Waterthrush, which I was able to document as defending territory as transients at Welder (see Chapter 5 on spring migration). However, I think that the majority of transient songbirds at Welder were, in fact, in a feeding state regardless of their brief stay. The reason that they were social rather than territorial was that food, in the form of arthropods, mainly insects, was superabundant, requiring no effort to sequester. The importance of this distinction, that is, whether they were just hanging out or actively trying to rebuild fat reserves, determines how food availability is considered as a factor in Alan's equations. If food is superabundant, then there is no carrying capacity for this factor in this habitat; if it is limited, then there is a carrying capacity, and some individuals will be forced to leave or die, and again, age or sex may be aspects that must be taken into account—which Alan's equations allow.

The fourth major collection of habitats for which carrying capacity equations must be considered are those occupied during the wintering period. Following our Hooded Warbler example, we know that these birds arrived on their wintering grounds in Tuxtla rain forest in October, establishing individual feeding territories that they defended until their departure northward in April. Thus, an important part of the equation for calculation of limiting determinants for annual population size of the Hooded Warbler is amount of lowland tropical rain forest in Middle America. However, we also know that male territories in that habitat outnumbered females eight to one. Female territories were found more commonly in a different habitat altogether, namely thickets. Therefore, we need to take into account carrying capacity for the different sexes in different wintering habitats. In addition, there are many Hoodeds, among which juveniles likely are disproportionately represented, that are unable to locate and defend a territory in any habitat and end up wandering though occupied territories and scrubby borders, perhaps throughout the entire wintering period. We know from Wood Thrush studies that mortality is likely much higher for these wanderers by at least an order of magnitude. Still, it is not one hundred percent. Some portion of these non-territory-holding birds survive to return to the breeding ground. Birds occupying this collection of wintering habitats for which there is no carrying capacity also have to be taken into account, which Dr. Pine's equation allows.

Thus we can see how the Pine equations allow for calculating which habitat over the course of the year exercises the principal control over population size for each age and sex category of the Hooded Warbler. As we have discussed, many species of long-distance migrant songbirds, such as Wood Thrush, Kentucky Warbler, Yellow-bellied Flycatcher, and so forth, seem to fit the Hooded model. For these, wintering-ground habitat limitations likely exert limits to population size. However, many other species of migrants do not fit the Hooded model, such as many waterfowl and shorebirds, whose winter habitats do not appear to offer density-dependent limits to population size. Of what use are Dr. Pine's equations under these circumstances? The value is that they force consideration of each part of the annual cycle in terms of its *potential* effects on population size for all age and sex groups of all migrant species, regardless of their mode of existence. If on close examination it is determined that a given species suffers no density-dependent limitations on any portion of its population at any point in the annual cycle, well that's fine. It means that the situation applies that was intimated by the letter to the editor of *Audubon Naturalist Society Newsletter* in response to my article offering habitat as the main concern for bird populations, namely that all factors combined affecting increase and decrease determine population size. This situation likely is true for most species of organisms living on the planet. It is only for longer-lived organisms that competition becomes a factor.

The extraordinary value of Dr. Pine's equations for migrants, or really any species that occupies different habitats during different parts of the annual or life cycle, is that they establish a model for a completely new way to think about the population ecology of the species. No longer was the migrant a prisoner of its breeding period. Now, its entire annual cycle was opened to inspection, leading the way toward a much clearer conception of what migration is.

CHAPTER 9

ORIGIN
and Evolution

> "There's no use trying," she said: "one *can't*
> believe impossible things."
> "I daresay you haven't had much practice," said
> the Queen. "When I was your age, I always did it for
> half-an-hour a day. Why, sometimes I've believed as
> many as six impossible things before breakfast."
> —Lewis Carroll, *Through the Looking-Glass and What Alice Found There*

According to many old wives, around the time Pocahontas and Captain John Smith were first establishing their complicated relationship (1607), "a squirrel could travel from the Atlantic to the Mississippi without touching the ground." Whether or not this adage is literally true, there was certainly a heck of a lot of forest covering the vast majority of the eastern half of the continent at the time of European settlement. This situation, however, changed very rapidly with colonization, and by the mid-1800s most of the primeval wilderness was gone, replaced by pasture, farmland, orchards, and woodlots of second-growth timber. New York and New England reached maximum forest loss by the 1840s.

You may be forgiven for wondering how this brief lesson in historical landcover is relevant to our theme. My purpose in its exposition is to explain how migration originates by presenting some data resulting from what we know actually happened as this extraordinary scenario of continent-wide habitat manipulation played out over half a millennium. Consider this catastrophic sequence as an experiment designed by a somewhat deranged Manipulator. The question under consideration would be, *How will the avifauna be affected by conversion of immense portions of the continent from forest to farmland over a 200–year period followed by a subsequent 150–year period where much of the land reverts to forest?*

Hypothesis formulation and testing are key parts of this issue of attempting to sort observations into meaningful explanatory constructs. Ronald Fisher, who laid much of the foundation for statistics, proposed that the base or "null" hypothesis should state the assumed situation that there is no relationship between two phenomena. A test is formulated on this basis, data are collected and subjected to a statistical analysis, and a probability is calculated for the likelihood that the null hypothesis is correct. If it is found to be incorrect, usually at a probability greater than ninety-five percent, then it is rejected, and a new, alternative hypothesis is formulated. That is the reasoning that I have tried to follow in my own research, and to present in this account. It seems backward, of course. Shouldn't you just state what you think you are going to find and not play any mental games? That approach is what they now expect in National Science Foundation proposals, or at least that was the case in the early 2000s. My own sense is that by stating what you think, you are going to find you lose some objectivity, which is always a precious commodity for a research scientist.

The null hypothesis for the first part of this experiment (forest removal) is as follows: *Creation of entirely new "open" habitats (pasture, farmland, and old field second growth) from forest* will have no effect *on the community of birds.*

Unfortunately, early information is mostly lacking. Explorers and pioneers in the 1600s and early 1700s seldom comment on wildlife in their notes, other than on danger posed or recipes. Nevertheless, some excellent work was being done by the early 1800s, which allows for sufficient informed speculation to reach some tentative conclusions.

The chief subject for our experiment is the Bachman's Sparrow, an inconspicuous little mite first described as a species based on a specimen collected by John Abbott near Savannah, Georgia, in the early 1800s. Abbott shot the bird in a habitat unique to the drier uplands of the coastal plain of the southeastern United States, the pine savanna; indeed, the vernacular name first bestowed was "Pine-woods Sparrow." This floral community has been depicted by some as "park-like," but, as anyone who has worked in it can tell you, that characterization is grossly inaccurate. While it is true that the immense long-leaf and loblolly pines of old growth savanna are widely spaced, attempting to make one's way through the understory of dense, waist-high palmetto and saw grass interspersed with impenetrable thickets of scrub oak, green briar, holly, and the like is no one's idea of "a walk in the park."

The Bachman's Sparrow, despite its undistinguished appearance, is the poster child for this habitat because it possesses one of the most wondrous songs in the avian repertoire (my personal opinion), a bubbling, sweet, reed-like flow of piercing purity and heart-breaking beauty, delivered in flight or from an exposed branch in spring and early summer.

At the time of its discovery, and indeed well into the 1800s, the Bachman's Sparrow was known only as a year-round resident of the pine savanna of the southeastern coastal plain, and described as such by John James Audubon in the 1830s based on first-hand experience. However, a century later, two populations existed—one resident in the southeastern coastal plain and a migratory form, known to breed from the foothills of the southeast, northward hundreds of miles to the fields and hedgerows of Illinois, Ohio, and Pennsylvania, and wintering along with its southern relatives in the southeastern pines.

Thus, the first two hundred years of our experiment, the forest-removal portion, completely altered the distribution of the Bachman's Sparrow, doubling its range and spawning a new population composed entirely of migratory members. But the experiment was not over. Subsistence farming, never an especially attractive career choice, began a marked decline after the Civil War as market forces (better agricultural land westward; better

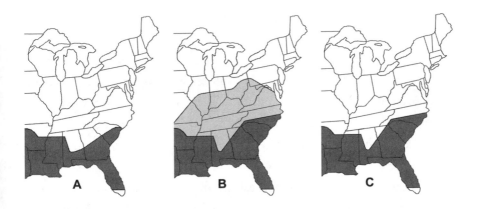

FIG. 9.1. Bachman's Sparrow range in eastern North America in 1800 (A), 1900 (B), and 1996 (C). Dark gray = permanent resident; light gray = summer resident. Summer residents migrated to "permanent resident" portion of range in winter.

income in cities) moved people off from hard-scrabble homesteads. As I write this section (December 10, 2018), the trees are back across many parts of eastern North America. What had been huge areas of grasslands and scrubby overgrown fields for a couple of centuries are now urban centers, industrial agriculture, and forest. This change has been quite beneficial for the warblers, vireos, thrushes, and so forth that use woodlands for breeding, but it has not been good for some open country species, such as our Bachman's Sparrow. The migratory population of this bird has disappeared from eastern North America, presumably due to lack of suitable scrubby habitat. Interestingly, the southern resident populations from which the migratory populations were derived are unchanged, at least where their habitats remain (Figure 9.1).

The experiment is still underway, but we can draw some conclusions based on what we have seen so far. Conversion of half the continent from forest to grassland and scrub during the early centuries of settlement resulted in huge changes in the avifauna, producing some new migrants from what had been southern resident populations (Bewick's Wren and Loggerhead Shrike in addition to Bachman's Sparrow) or allowing enormous range expansion for several others (Northern Bobwhite Quail, Eastern Meadowlark, Brown-headed Cowbird, Field Sparrow, etc.). Then, reversion of immense amounts of scrub to forest in subsequent centuries up to the present was followed by additional dramatic changes in the avifauna, including disappearance of the northeastern migratory populations of birds that had been present solely as residents in the southeastern United States at the time of colonization.

Thus, we reject the null hypothesis predicting no effect, and we formulate a new one as follows: *Creation of entirely new habitats* has profound effects on the community of birds, *causing many species to massively expand their ranges. Those species able to find food year-round in the new habitat become permanent residents, whereas those able to find food seasonally occupy the new habitat as migrants. Disappearance of these habitats causes reversion to the way things were.*

Perhaps all of this seems obvious, as so much does when pointed out. But in fact, it is not obvious to most, and it raises a number of questions to which we do not presently know the answers. For the purposes of our discussion, the critical questions are *from whence and how did the Yankee population of* migratory *Bachman's Sparrows originate?*

The only possible answer to the first question is *from the resident population of sparrows in the southeastern pine savannas.* There is also only one possible answer to the second question, namely, *dispersal from the original southeastern pinelands population.* Young birds dispersing from the southeastern parent population discovered the vast newly created scrubby farmlands to the north left in the wake of New World colonization. Food in these fields was superabundant for the sparrows during the spring and summer months, allowing them to settle, breed, and raise their offspring. Had these immigrants been able to find sufficient food throughout the year, these recently invaded lands would have become part of a huge range expansion, as was true for another southern savanna and grassland species, the Northern Bobwhite Quail, which quickly became a staple on the settler menu as its range as a resident species spread northward all the way up into

Canada within decades of the massive deforestation associated with westward settlement that took place after the close of the Revolutionary War. However, dispersing Bachman's Sparrows found habitat that could only be exploited seasonally—suitable, and in fact, superlative, during the breeding season, but where food resources became scarce or absent later in the year. As a result, they became migrants, returning to the parent population for the winter months and departing from it again the following spring when members of the resident population begin to establish and defend their breeding territories.

I recognize that the highly speculative account related above sounds fanciful, including as it does a number of questions for which we currently have no good answers. But remember, it is speculation based on some incontrovertible facts, to wit: (1) the Bachman's Sparrow was an indigenous species whose world range at the time of European arrival in North America was restricted to pine savanna of the southeastern coastal plain; (2) conversion of much of the eastern half of the continent from forest to farmland and old fields was accompanied by the appearance of a migratory population of the sparrow within a matter of decades; and (3) reversion of farmland to forest was accompanied by the disappearance of this migratory population, also within a matter of decades.

It is my contention that these facts explain the origin of migration for every animal species. The origin of migration can be stated as follows: *Migration originates when members of a resident population disperse to a new seasonal environment suitable for breeding but unsuitable for year-round residence, requiring return by the dispersing individuals to their population of origin once breeding has been completed.*

Evidence of actual appearance and disappearance of migration in Bachman's Sparrow and several other species, as well as the history of post-Pleistocene occupation of high-latitude habitats and the now well-documented phenomenon of intratropical migration in significant numbers of South American, African, and Asian species, further confirms the ability of resident birds to exploit seasonal environments by developing migration even in the *absence* of seasonal change in their birth places.

Exaptations versus Adaptations
In which a structure or behavior useful in
one evolutionary context becomes co-opted for another

Migration obviously requires a special suite of characters in order for the individual to survive and reproduce. Where did these characters come from if not from prolonged exposure to natural selection through some process such as that suggested by Northern Home or Migratory Syndrome theorists? To answer this question, I think that an idea initially put forward by Darwin, and developed further by Stephen Jay Gould and Elizabeth Vrba, is helpful. Gould and Vrba suggested that many of the characters considered necessary to conduct a particular activity evolved originally through the process of natural selection favoring successful performance of a different activity. Gould and Vrba termed the characters evolved for the primitive activity "exaptations," whereas the characters derived from natural selection on the exaptation for the new activity they termed "adaptations." The eminent Dutch ornithologist Theunis Piersma and colleagues famously applied the concept to an analysis of migratory traits in a 2005 paper to which much of the discussion below owes a debt of gratitude. (Actually, I think a discussion over coffee at a meeting in the early 2000s with Barbara Helm was the inspiration for me.)

This terminology is admittedly a bit confusing, mostly due to the fact that the term "adaptation" is in general used to refer to both types of evolutionary change. The reason Gould and Vrba chose these particular words was to make an important distinction between two different types of adaptations while avoiding any intimation of teleology, that is, suggesting that a species evolves a character because it needs it, as proposed by Lamarck for the long neck of the giraffe and similar remarkable evident compliances between structure and environment.

An example that is often used to explain the distinction between "exaptation" and "adaptation" is the feathers of birds. Feathers may originally have evolved to enhance thermoregulation or display, but as the outer covering for the forearm they became co-opted into primaries and secondaries through natural selection favoring enhancement of flight. Thus feathers may be an *exaptation* for thermoregulation but flight feathers are an *adaptation* (sensu Gould and Vrba) for flight. This example also illustrates how the whole enterprise can get a bit subjective when examining the various traits required for an activity two or three steps removed from their original purpose. Flight feathers, for instance, are an *exaptation* for flight, but the especially elongated flight feathers of the Eskimo Curlew appear to be an *adaptation* for long-distance migration.

Despite this element of potential muddle, I think that the distinction made by Gould and Vrba can help us understand how migration can originate in a single generation of nonmigratory birds by considering most of the characters required for successful migration (in a fitness sense), such as flight, fattening, navigation, homing, and time-keeping, as *exaptations* for successful dispersal. Indeed, all of these characters have been found through field observation or experimentation to be in many species of nonmigratory birds, as explained in detail in my earlier book *The Avian Migrant*.

Perhaps the best illustration of this principle is the *sine qua non* of adaptations for migration, *Zugstimmung* (migratory flight), or its stand-in in experimental studies with caged birds, *Zugunruhe* (frustrated migratory flight). First described in the early 1700s, the experimental work resulting in the naming of the behavior was done in the 1920s by the German scientist Franz Groebbels. As discussed in Chapter 3, Groebbels described two different physiological states associated with migration: *Zugdisposition*, during which the bird ate intensively and laid down reserves of subcutaneous fat for a week or so prior to departure, and *Zugstimmung*, when the bird commenced actual migratory flight. If the bird was prevented from leaving by being held in a cage, it remained active, jumping around and/or repeatedly attempting to launch itself into the air. This behavior, termed *Zugunruhe*, often continued for several hours during the night, and for weeks or even months during the migratory period.

Beginning in the 1950s, German scientists began to build on the work of Groebbels through a series of experiments with caged birds in which

Zugunruhe was measured in a number of different migratory species under different experimental conditions. These kinds of studies continue to the present day and have resulted in remarkable revelations regarding interactions between genes and the environment concerning control over the migratory habit.

As mentioned in Chapter 3, blackcap warblers from the Cape Verde Islands show *Zugunruhe*, despite the fact that this is not a migratory population. Dr. Eberhard Gwinner found this observation intriguing and encouraged his brilliant PhD student, Barbara Helm, to focus on factors affecting *Zugunruhe* in populations of different migratory status of the same species. The bird chosen for this work was the stonechat, which has long-distant migrant populations that breed in Siberia and winter in South Asia; short-distance migrants that breed in Austria and winter in the Mediterranean region; partially migratory populations, some members of which breed and winter in Ireland, while others breed in Ireland and winter in the Mediterranean; and tropical nonmigratory populations in Africa and South Asia.

Dr. Helm's studies have been underway for more than two decades now, and they have yielded a number of extraordinary insights, presented in a series of seminal publications. As a warning to the reader, it is best to read the original work, not only because I may inadvertently misrepresent the extensive amount of data presented but also because, unlike Dr. Helm, I bring a particular bias to their interpretation, namely my conviction that migration is a form of dispersal. A summary of the findings I deem relevant to our discussion of *Zugunruhe* as an exaptation for migration is presented below, along with my own comments regarding how these contribute to our understanding of the relationship between migration and dispersal.

1. *First, a warning:* Zugunruhe *is not* Zugstimmung. *Zugunruhe* is a behavior assumed to result from frustration of *Zugstimmung* by being held in a cage. A finding that other researchers have noted is that while *Zugunruhe* appears to last throughout the weeks or months comprising the migratory season for the species, free-flying birds undergo relatively few nights of actual flight, alternating with stops of varying length to rest, refuel, and/or

await appropriate weather conditions favoring movement. Relatively few studies have been done to test what aspects of *Zugunruhe* accurately reflect movement and what are artifacts of the caged-in situation, and they are inconclusive.

2. *Older birds (at least one year old) demonstrate far less* Zugunruhe *than juveniles.* If *Zugunruhe* were simply a reflection of the migratory journey, age seemingly would have little effect on its duration since adults and immatures are assumed to follow the same genetic program for timing and destination. However, if the behavior is reflective of attempted movement associated with dispersal, of which migration is only a part, then age differences in continuance would be expected for at least three reasons. First, because young birds on the breeding ground, unlike their parents, would be unfamiliar with location of potential habitats in which to undergo molt and preparation for migration, and they would have to search for them if the breeding territory and its immediate environs were not suitable (or were defended), as is often the case as discussed in Chapter 7. Second, juveniles would be novices in terms of locating stopover areas during migration, thus requiring some search for routes appropriate to their experience (or lack thereof). Third, and most importantly, young birds would have no experience in locating appropriate wintering habitats or territories, and thus they would be expected to have to continue dispersal on the wintering ground, possibly long after their arrival. Indeed, as discussed in Chapter 4, such searching could last throughout the entire wintering period.

3. *Juvenile birds show* Zugunruhe *long before the migratory period for stonechats begins, actually even before or during the molting period.* Once juveniles achieve independence from their parents, usually three weeks or so after fledging in songbirds, they are on their own. They must find an undefended appropriate piece of habitat with sufficient food to allow for molt and preparation for migration. In addition, they need to familiarize themselves with the surrounding area in terms of potential sites for breeding when they return in spring. These needs may require movement over some distance, for which physiological preparations (*Zugdisposition*) followed by actual movement (*Zugstimmung*) would be entirely appropriate.

4. *Tropical nonmigratory Kenyan stonechats show* Zugunruhe *in both fall and spring.* Nocturnal activity in caged individuals at the end of the breeding period has been reported for several nonmigratory species, including, as mentioned in Chapter 3, Cape Verde blackcaps. Bob Zink and other molecular systematists supportive of the Migratory Syndrome theory state that this behavior has no adaptive function—that it is, in fact, a result of "incomplete suppression" of a trait once useful in eons past when the species was migratory. There are two assumptions involved in this assertion: (1) the nonmigratory population in which the behavior is observed was once migratory, and (2) the trait currently has no selective value. There are no data to support either of these assumptions. I have addressed elsewhere the first assumption of temperate zone origin for tropical residents with migratory relatives. The second assumption, that tropical stonechats are saddled with a useless behavior—one that evolved one to three million years in the past (the time estimated by molecular geneticists that the African population has been separate from migrant stonechats)—that consumes both time and energy, can be dealt with, I think, by elucidating the possible fitness advantages to a resident bird. These derive from principles presented in Chapter 8, namely that in a stable, density-dependent population, offspring are likely to be forced by their parents to move away (disperse) from their natal area once they have reached independence. I propose that fattening in preparation for this movement (*Zugdisposition*) as well as the actual movement itself (*Zugstimmung*) would be highly advantageous in a fitness sense for individuals possessing the capability.

As I have mentioned, several studies have reported fall *Zugunruhe* in several nonmigratory species. Spring *Zugunruhe* has been less-commonly reported. Nevertheless, capability for biannual movement as a normal part of dispersal strategy makes excellent sense from a fitness perspective. Assume that the bird disperses initially when forced out by competitors and locates a suitable seasonal environment for survival. If that habitat becomes unsuitable during a later season, then ability to return to the place from which it originated would have considerable value.

5. *Changes in environmental conditions (sunlight, rainfall, food avail-*

ability) can cause Zugunruhe *to begin regardless of the season.* As proposed above, the physiological and behavioral traits normally associated with migration could have value in a variety of circumstances in which social or environmental situations alter the availability of a critical resource making movement required.

Dispersal and Migration

The bottom line for me is, if you can move, you can migrate, given the appropriate environmental conditions. This idea is well substantiated by the types of organisms known to migrate, which, in addition to birds, include fish, butterflies, grasshoppers, squid, caribou, whales, bats, turtles, salamanders, frogs, and many, many others.

I have placed emphasis on this concept because it is the reason for the existence of this book (and for *The Avian Migrant,* for that matter). I have attempted in earlier chapters to build an understanding of how this Dispersal Theory for the origin of migration is inescapable, despite being entirely contrary to the accepted paradigm of migrant appearance, which requires genetic change in a resident population resulting from selection pressure imposed by gradual disappearance of food resources during the nonbreeding seasons as a result of decades- or centuries-long climate change. This hypothesis, of course, is often referred to as the Northern Home Theory or Migratory Syndrome theory for migrant evolution.

I readily admit that the idea that all organisms that can move already possess the suite of characters necessary for migration sounds a bit like the Migratory Syndrome theory in which generations of natural selection produce a genetic package or set of traits necessary for successful migration—a package that can be turned on and off by a single genetic switch. According to this theory, a resident population can possess the characters without using them as a result of evolution of migration during some previous period of environmental pressures millions of years in the species' ancestral past. When the selective environment changes from less seasonal to more seasonal, natural selection will then favor the expression of these characters by turning on the genetic switch. This turning on of the switch would not be immediate, of course, but one could imagine it taking place over a few generations if the selective pressure (such as lack of food as a result of cold weather) were sufficiently intense.

Despite superficial similarities, the Dispersal Theory for origin of migration differs radically from the Migratory Syndrome theory in at least four essential elements:

- *With the Dispersal Theory, no genetic change in the population is required for movement to occur. Successful dispersal has selective value for every organism that is a member of a density-dependent population and migration is simply a form of dispersal.*

- *No change in the climate in the range of the originating population is required to force initial movement, so there is no need to posit a genetic switch with the Dispersal Theory. The pressure for dispersal comes from competitors of the disperser's own species.*

- *According to the Dispersal Theory, change in the weather is not the ultimate cause for return of migrants from breeding area to population of origin. That behavior comes about due to the lack of food during part of the annual cycle, which results from the seasonal nature of the breeding habitat. Since the disperser is native to a place where food resources are present year-round, it can return to that place once breeding has been completed and food resources diminish.*

- *The Migratory Syndrome theory requires annual movement* away *from the environment in which the organism has a niche (the breeding ground)* and *into* a new environment (the wintering ground) where it does not. In order for this strategy to work in a fitness sense, there must be superabundant resources on the wintering ground or a lack of competitors. Absent these conditions, the wandering migrant would not be able to survive the nonbreeding period in places where it could not compete for critical food resources. *In contrast, according to the Dispersal Theory, the disperser* moves into *an environment where resources are superabundant for breeding and* returns *to its native area, where it has a niche, when breeding-ground resources decline as a result of annual climate change, like our Yankee population of Bachman's Sparrows.*

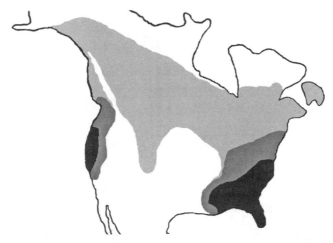

FIG. 9.2. Extent of forest distribution in North America at present (*light gray*), 12,000 years ago (*dark gray*), and 18,000 years ago (*black*).
BASED ON WOODING AND WARD 1997

Now, you may say that the Dispersal Theory for the origin of migration is quite a load to put on the back of one little sparrow. And you would be right, except for the fact that Bachman's Sparrow is a representative, a stand-in, for the entire class of migratory organisms. What is unique about this sparrow is that key elements of the process have occurred literally almost before our eyes, in real time. For other migrants, we have to do a bit more creative thinking. But the same picture appears once the effort has been made.

Another example is represented by those species that migrate each spring from the tropics to the temperate and boreal forests of the Northern Hemisphere. During the most recent glacial maximum, roughly seventeen thousand years ago, modern breeding ranges in North America for our migrants were covered by ice, tundra, and cold deserts. Only two major pockets of forest existed: one was in the southwestern region; the other was in the southeast (Figure 9.2). Temperate deciduous and boreal forests likely did not reach their current range until just a few thousand years ago. Thus, for these hundreds of migrant forest-related species, we are left with the same questions confronting us as applied to the Bachman's Sparrow populations of the northeastern United States: *from whence and how did these long-distance migrants from the tropics come to occupy breeding habitats in the high latitudes of the continent in a few thousand years?* We know one thing for certain: they did not originate as permanent residents pushed south gradually over centuries by incremental changes in winter weather. Their current breeding ranges did not exist a few thousand years ago—and had not existed for one hundred thousand years or so since the previous interglacial period. Furthermore, when these high-latitude forests did appear, it was as a result of warming, not cooling.

The answers to these questions are the same for the entire community of temperate and boreal migrants as they are for the Bachman's Sparrow: they originated from breeding populations from more southerly latitudes, and the process by which they came was simply that of normal postbreeding-period dispersal. What we do not know is whether these dispersers derived from (1) populations that were resident in the tropics, (2) populations resident in southern temperate refugia, or (3) migrant breeding populations in those refugia that wintered in the tropics.

Derivation of the colonizers for newly developing habitats in boreal regions and north temperate regions would depend upon the ecology for each of the species concerned. Our understanding of species composition of the flora and fauna of temperate zone refugia is still at an early stage, but it is unlikely that they contained sufficient diversity to support all of the habitat requirements for several hundred species of migrants. Thus the probability is that dispersers for some north-colonizing migrants derived from migrant breeding populations in the temperate zone refugia, while others derived from year-round residents in the refugia, and still others came from populations resident in the tropics.

The fact that bad weather provides an obvious evolutionary incentive for migrants to head south makes the Northern Home Theory easy to understand. Similarly, the lack of bad weather in the tropics has comparable power, as clearly expressed by Ian Newton in his magisterial treatment of bird migration, *The Migration Ecology of Birds*, where he observes, "In contrast [to those living in higher latitudes], birds confined to lowland equatorial rainforest are probably the least migratory, especially the small insectivores of the understorey where conditions remain relatively stable and suitable year-round. This year-round consistency in the rainforest environment removes any advantage in moving, and many individuals may remain within the same few hectares throughout their adult lives."

This statement neatly summarizes the main case for both the Northern Home and Migratory Syndrome theories, namely, that without some seasonal change in one's home environment, there is no need to migrate. This idea is, of course, in direct opposition to that which forms the central theme of this book, which is that the ability to disperse from the natal area is favored for members of *all* species, whether they live in a seasonal environment or not, and those that can disperse successfully will do so, even if it requires subsequent return (migration) to their point of origin at a later date. The explanation for the relatively small number of species known to migrate from equatorial rain forest into seasonal environments is that for most of these species their niches are too specialized to allow them to move long distances successfully (in a fitness sense) to seasonal environments for breeding. Proof of this postulate is the fact that many birds living in relatively stable tropical habitats, including lowland rain forest, *do* migrate to seasonal environments to breed. In fact, intratropical migration provides solid support for the dispersal hypothesis for migration origin. Although less prevalent among tropical communities as a whole when compared with those of temperate and boreal areas, seasonal movements within the tropics are much more common than previously thought. The chief reason for our lack of understanding regarding the phenomenon has been a dearth of knowledge. There are no continent-wide banding programs for the tropics as there are for North America (indeed, we are unique in the world in that regard), and without such an enterprise, data on population movement is difficult to come by. Nevertheless, long-term studies by several tropical field biologists have provided excellent

Homing

Eros is not the only near-universal attribute in the animal kingdom; homing (*ortstreue*, philopatry, site fidelity) shares that ecumenical aspect as well. This fact has been well documented in an extraordinary diversity of organisms as summarized in comprehensive reviews of the topic by Roswitha Wiltschko, Hugh Dingle, Reginald Baker, and several others. Defined as the ability to return to point of origin, homing has been documented in mammals, reptiles, amphibians, fish, insects, squid, and birds, of course—in effect in nearly every group of mobile organisms in which it has been investigated (displacement experiments).

Navigation and orientation constitute the nuts and bolts of the homing phenomenon, the "how" as it were. Ingenious experiments, conducted mainly in the past three or four decades, have demonstrated that nearly every aspect of the environment that could be detected and useful for an organism attempting to orient itself in order to move back to where it came from is used. In the process, researchers have found that not only all five of the known senses (touch, taste, hearing, smell, and sight) are employed by one or more species known to demonstrate homing ability but a wide array of other senses previously not known to exist are used as well (including detection of the earth's magnetic field, polarized light, infrared, ultraviolet, infrasound, star maps, the sun's orientation, and, of course, the ability to remember topographic features).

The variety of organisms in which homing is found and the multiplicity of sensory capabilities employed make the point quite clear that homing is as old as movement. Thus, I claim the behavior as an exaptation for migration.

Many researchers have proposed that the homing involved in the everyday wanderings in search of food or mates with subsequent return to territory or roosting spot are distinct from migration. However, in terms of the sensory capabilities required, this may be a distinction without a difference. In this regard, it is interesting to note that the Wiltschkos (Roswitha and Wolfgang), who pioneered much of the investigation of navigation and orientation, used the domestic pigeon, a wide-ranging but generally nonmigratory species, for their work. Basically, the only differences between daily "out and back" movements and migration, as far as the individual is concerned, are distance and duration. The same sensory capabilities are required for both, which explains why homing is as old as movement, and why we see the extraordinary range of sensory capabilities enhancing its success in such a wide variety of animals.

There is, however, one very large difference between the homing involved in daily movements and the homing involved in migration—the young of the original migrants are capable of performing it. This feat is the core mystery of migration—not the fact that an individual bird can travel thousands of miles to return to a place where *it has been,* but that its *offspring* can cross oceans and continents *to get to a place where they have never been.* The behavioral act of homing is the most important preadaptation for migration, and it has to be inherited—passed on from parents to young. Weirdly, there is evidence that it is a fundamental property of most or all organisms that are capable of unaided movement.

The Germans, once again, have led the way in tackling this inscrutable problem, at least in birds (bats, butterflies, and probably many other migratory groups likely have the same ability). Peter Berthold and Eberhard Gwinner were the chief architects of these studies, which have yielded a remarkable amount of information about how novice birds might find their way from breeding grounds in Germany where they were hatched to wintering grounds in Africa.

Chapter 3 describes the set of experiments used by Berthold and his colleagues in Radolfzell, Germany, to demonstrate the genetically programmed nature of *Zugunruhe* (frustrated movement) in terms of both direction and duration in a long-distance migrant Old World warbler, the blackcap. In these experiments, juveniles from three different populations of blackcaps were tested: (1) birds from a population that breeds in western Europe and winters in western Africa; (2) birds from a population that breeds in eastern Europe and winters in eastern Africa; and (3) birds resident in the tropical Cape Verde Islands. Young birds from each of the different populations oriented in a different way: western European birds to the southwest, toward their western African wintering grounds; eastern European birds to the southeast, toward their eastern African wintering grounds; and, incredibly, nonmigratory Cape Verde birds toward that Archipelago. In addition, duration of *Zugunruhe* differed between the populations in a pattern seemingly appropriate to the distance to be migrated.

Based on these results, the researchers developed an explanation for how juveniles find their way from breeding to wintering areas based on a genetic program containing information on direction of the wintering area and the amount of time required to get there. The Wiltschkos described this hypothesis as follows in Berthold et al.'s *Avian Migration* (pp. 433–456):

> Inexperienced young migrants are guided by innate information on their migration route given in polar coordinates, namely as a distance and a direction from the starting point. . . . An endogenous program that is genetically transmitted from one generation to the next . . . determines the direction and approximate length of the route, the latter by controlling the duration of migratory activity. . . . Put into human terms, the innate information would be equivalent to an instruction like: "fly for 6 weeks towards the southwest," or, in cases of nonstraight routes like e.g. [*sic*] those of the garden warblers, *Sylvia borin*, from the southern German population . . . to something like: "fly for 6 weeks towards the southwest and then for 7 weeks towards the south-southeast."

This ingenious theory, known variously as "clock and compass," "migratory route," "bearing and distance," "vector," and probably some other names in the literature, is currently the predominant hypothesis explaining how young migrants find their way to distant ancestral wintering areas. There are, however, some problems.

1. *Zugunruhe* is not migratory flight. It is a presumed frustration of that activity.

2. While *Zugunruhe* duration demonstrates some relationship to migration distance, it is not a "tight correlation," as has often been claimed. Indeed, there is no reason to expect a close relationship between *Zugunruhe* duration and migration distance given the likely possibility of drastically different situations encountered by different individual migrants (weather, food availability, competition, and so forth).

3. *Zugunruhe* often lasts months longer than the migration period.

4. *Zugunruhe* can occur spontaneously outside the migration season in seeming response to variations in environmental conditions (food availability, weather).

5. Migration routes are not fixed. Northern routes are often different from routes taken south, and young birds often follow different routes from adults.

6. Individuals (siblings) with presumably the same genetic programs can follow different migration programs.

7. The same individual can follow one migration program in one year and a different program the next.

Despite these quibbles, the majority of the scientific community holds tightly to the Migration Route hypothesis as the best explanation we have. Molecular systematists in particular have found it a congenial base from which to construct a variety of imaginative conjectures. Be that as it may, even those who accept that the Migratory Route hypothesis might be flawed can be forgiven for demanding, "*So what's the alternative?*" I propose that the Radolfzell experiments suggest one.

One of the most remarkable findings from these experiments, for me at least, was the discovery of *Zugunruhe* in the Cape Verde birds. Not only did these inexperienced young birds show restlessness (as though they were migrants), they also *oriented properly in the direction of these islands* from their German laboratory home.

Although Berthold does not say so, I have to believe that this behavior must have come as a shock to the researchers. They probably included the Cape Verde birds as a control—that is, as a nonmigratory population in which the experimental subjects would be expected to show no migratory behavior whatsoever in contrast with birds from the two migratory populations. When the birds did show migratory behavior, the researchers must have had to do some pretty creative thinking. The explanation that they came up with was that migratory behavior in these tropical residents was an atavism (like an appendix?). They suggested that the Cape Verde population likely was recently (thousands of years? millions?) derived from a European migratory population that simply stayed on to breed. Thus, the migratory behavior shown by the young birds was just a sort of genetic ghost passed on from their migratory ancestors.

But consider the implications. Do juvenile blackcaps raised by their parents in the Cape Verdes start hopping around as autumn comes (if there is "autumn")? If not, is *Zugunruhe* simply an artifact of being held in a cage? These noodlings are forced on us by the surmise that migratory behavior, including the route program, takes many generations to evolve. According to this notion, such behavior in a member of a tropical resident population of blackcaps has to be a genetic remnant from a time when the population was migratory. What other explanation could there possibly be? But what if we assume that the Cape Verde population was never migratory? Is there any way that this could help to explain the results from the German experiments? I suggest that there is. What if it is the *parent's point of origin* that is somehow programmed into the young bird (as appears to be the case for monarch butterflies)? If that were true, then the behavior of the Cape Verde subjects fits perfectly with those from the European-breeding populations; all three are attempting to home to their parents' wintering areas. I suggest that this would be pretty easy to test, if you had the fantastic logistics of Radolfzell (which almost no one else does). Just take a few juveniles of other tropical resident species to your temperate zone lab and see how they behave when fall rolls around. Will they "know" that they are not where they are supposed to be and make an effort to try to get to their tropical home?

Sir Arthur Landsborough Thomson suggested in his book *Problems of Bird-Migration* (1926) that migration *destination* might somehow be genetically programmed. Rice University professor George Williams also suggested this idea in his Lida Scott Brown lectures in 1958, and there is extensive evidence to indicate that it is possible. More than eighty experiments have been performed in which young birds captured and displaced hundreds of miles from their breeding site were nonetheless able to orient correctly to their wintering area.

A striking example of a natural test of the Migratory Route hypothesis is the range expansion of the Northern Wheatear (see Figure 3.5). This bird, an Old World native that breeds across northern Eurasia and winters in central Africa, began to expand its breeding range into the New World over a century ago, occupying Alaska and neighboring Arctic Canada in the west and Baffin Island, Arctic Quebec, and Labrador in the east. If the New World colonists were dependent upon some ancient Migratory Route program to make their way to the wintering ground, the eastern birds would have ended up in the middle of the Atlantic Ocean, while the western birds would be wintering in the Pacific. However, a destination program would have them traveling to central Africa for the winter, which is what they have been doing quite successfully for the past century.

The work of one of Berthold's students, the late Andreas Helbig, seemingly provided critical support for Migration Route hypothesis. Helbig bred blackcaps from western Europe that wintered in western Africa (and hence showed southwestern *Zugunruhe* orientation) with birds from eastern Europe that wintered in eastern Africa (showing southeastern *Zugunruhe* orientation). The hybrids showed southern *Zugunruhe* orientation. This finding was a eureka moment, appearing to confirm a genetically based Migratory Syndrome. Molecular systematists found the discovery especially exciting as it could provide an explanation for reduced fitness of hybrids between populations that followed different migration routes (that is, they assumed the hybrids followed intermediate routes which took them to places where probability of surviving the winter was lower). Several such instances of "heterozygote disadvantage" assumed to result from mixture of Migratory Route programs have now been published. However, in none of these was *Zugunruhe* of the two parent populations or the offspring tested. Lack of this corroboration is a potential problem. Barbara

Helm and her coworkers have found no such clear-cut mixing of *Zugunruhe* in their experiments with stonechats from populations that differ in wintering destination. In addition, simply assuming that Migratory Route program mixture causes hybrid fitness depression is circular, precluding other possible explanations, among which might be that mixture of destination programs resulting in wintering in inappropriate regions could also affect hybrid fitness.

The mechanism for homing is the Higgs boson of biological behavioral studies. It has to exist, but we don't have any idea how. Someday, we will find it.

Weather

> North Wind, North Wind, fierce in feature,
> You are still my fellow creature;
> Blow your worst, you can't freeze me;
> I fear you not, and so I am free!
>
> —"Shingebiss," as told by Olive Beaupré Miller
> in *My Book House: Story Time*

In the Ojibwe tale "Shingebiss," Little Brown Duck Shingebiss (an American Black Duck perhaps?) refuses to leave his boreal home on the shores of Lake Huron despite the approach of, and eventual seizure of the land by, the dreaded North Wind. As the season progresses and the days shorten, Shingebiss departs each morning from his hearth to find a small pool or seep where he can feed. Each night, the North Wind seeks the pool out and freezes it. Days pass into weeks, and weeks into months, as Shingebiss follows the same daily routine, locating and exploiting a new, unfrozen spot. Eventually the North Wind takes Shingebiss's insouciance personally, threatening immediate death if the impudent dabbler continues his defiance by refusing to fly south. Shingebiss replies with the refrain cited above, driving the North Wind into a frenzy from which the North Wind emerges drastically weakened. The days are lengthening. The sun is returning. Spring has come. Courage born of knowledge has given Shingebiss the win.

For my purposes, this story carries an additional message, namely that weather does have an effect on movement for some kinds of birds that use some kinds of resources. For these species, weather is a modifier of movement, a proximate cause. However, it is not an ultimate cause. That designation goes to effects of climate on resource availability, mainly food. In this context, it is important to remember that Shingebiss, after all, is alone. All his buddies, including Mrs. Shingebiss and the kids, left at the first signs of winter; maybe even before the first signs, back in October. As Homer and many an author since have noted, weather is the most obvious generator of movement. The question for us is, "How does weather relate to migration?" In some instances, it is, prima facie, the direct cause of southward movement. Any duck hunter or north-country birder knows that.

For reasons explained in several places earlier in this book, the preponderance of evidence argues against weather as an initiator of the migration phenomenon. New environments, created by climate change or other factors (such as human land use), play that role. Nevertheless, weather has an undoubted effect on the southward movement of many migratory species. In fact, as I have described earlier, the entire group of species known to migrate has been divided into two cohorts based on whether their movements seem to be largely determined by a genetic program for timing ("calendar" or "obligate" migrants) as opposed to those in which ambient weather clearly plays a role ("weather" or "facultative" migrants). Timing of migration for calendar migrants appears to be under the control of a genetic program based on photoperiod whereas weather migrants often seem to respond to current weather conditions for beginning their southward movements. Quick consideration of the foods on which the two groups depend makes the departure differences in timing between them readily understandable from an evolutionary perspective. The majority of calendar migrants depend upon flying insects for a major portion of their diet during the breeding season. For these birds, waiting until a cold front destroys their food base could be disastrous. They might have to move hundreds of miles before being able to find a place where their foods are readily available. Birds like ducks and swans, however, often need only move a short distance to find open water with suitable resources.

In essence, many altitudinal migrants in the tropics are also "weather" migrants. My long-time collaborator Mario Ramos and I learned this during our banding work in the Tuxtla Mountains back in the '70s and '80s. In fact, study of these movements constituted a significant part of Mario's PhD thesis. We found that individuals of species normally associated with cloud forest of higher elevations would show up in our nets in lowland rain forest during *nortes* (cold fronts with strong north winds, driving rain, and cool temperatures); these species included the White-necked Thrush, Slate-colored Solitaire, and Common Chlorospingus. These birds disappeared once the fronts had passed, presumably returning to their highland homes when the weather had ameliorated. Some winter-resident migrants also behaved in this way. Wood Thrushes were a common member of the lowland forest community, where each individual defended a territory throughout the winter months, as described in detail in earlier chapters. However, during *nortes*, many additional Wood Thrushes would show up on the territories of our winter-resident birds, remaining for a day or two in loose flocks, generally ignored by the territory owner, before presumably returning to their territories at the upper elevational edge of forest habitat suitable for Wood Thrushes.

We had a chance to see the process from the other end during our expedition into the Tuxtla highlands to the Santa Martha volcano crater in March of 1985. Kevin Winker and Steve Stucker worked on capturing Wood Thrush territory owners and equipping them with bands and radio transmitters. At about three thousand feet, we were working at the upper edge of Wood Thrush winter distribution in the Tuxtlas. We found that these territory owners left the area during *nortes*, presumably moving downslope, returning to their territories once the *norte* had passed. Clearly weather plays a role in movement, but it is not an ultimate cause of migratory behavior in a population.

Natural Selection

Survival of the fittest.

—Herbert Spencer, *Principles of Biology*

Evolution, so far as is known, occurs only through the process of natural selection, which is the effect of the environment on the reproductive success (fitness) of an individual. Characteristics or traits ("phenotype") are the observable expression of a particular genetic makeup. Different expressions of a character, such as brown eyes versus blue eyes, arise through a process of random mutation (change in gene structure). Each such mutation has a different "selective value" to its possessor in terms of survival and reproduction (fitness). Different expressions of a character often have different selective values in different environments.

I have proposed that no evolution is required for assumption of a migratory habit. All that's required is a reason to move (competition? environmental change?), an ability to move, and a reachable place to move to. Yet, once members of a population have adopted migration as a life history strategy, the process of evolution of structures, physiological attributes, and behaviors used in that activity that affect fitness of their owners through natural selection is underway. The aforementioned elongation of flight feathers in shorebirds that are long-distance migrants may be one example. A migratory habit is likely to have some effect on most aspects of an organism's life history, and the longer the time period during which a population has followed this path, the more profound those effects are likely to be. But what does "longer" mean in this context?

The question of the time required for change in characters or behaviors of migratory populations is fraught. Geneticists have calculated the span needed for genetic differences to accumulate between populations of the same species that are isolated from each other, and, assuming a process of natural selection acting on random mutations, they find that this process often requires tens of thousands or even millions of years for significant deviation between populations to occur. Furthermore, they state that in order for any new trait arising from such mutation to become distinctive for any given population, it must be not only physically (in a geographic sense) but genetically isolated (no interbreeding) from other populations.

Be that as it may, there are many examples in which profound changes in phenotype of migrant populations occur much more rapidly than this—over decades or years—even among those populations that are not completely isolated (in a genetic sense). How can this be? I suggest some possibilities below. Consider the case of our prototype, the Bachman's Sparrow. As mentioned earlier, there were no members of this species found north of the Carolinas in the early 1800s, yet by the end of that century there was a migratory breeding population found as far north as New Jersey, Pennsylvania, and Ohio. I have explained this remarkable development as a natural result of the creation of an entirely new seasonal scrub habitat by subsistence farmers that our birds could capitalize on through dispersal, requiring no genetic change. However, once this migratory population existed, genetic change did occur, and very rapidly, as evidenced by the fact that by the early 1900s ornithologists were able to recognize members of the two populations as separate subspecies *on the basis of appearance*. In fact, the northern migratory population was described as a separate subspecies (*Aimophila aestivalis bachmani*) from the southern nonmigratory population (*Aimophila aestivalis aestivalis*) by the American Ornithologists' Union in the *Check-list of North American Birds* published in 1931. A brief stop at the Smithsonian's national bird collection will allow the visitor to confirm these differences with their own eyes.

Sexual Selection

She likes me best!
—Conversation overheard at
Euclid Avenue School playground, circa 1955

Genetic change supposedly takes time. A lot of time. Bob Zink and John Klicka, in their famous paper in *Science*, estimate that mitochondrial DNA evolves at a clock-like rate of two percent per million years. Nevertheless, it appears that change can occur at a much more rapid rate under certain conditions, as demonstrated not only in the Bachman's Sparrow populations discussed above but also in eastern US populations of the Loggerhead Shrike, which developed a southeastern resident subspecies (*Lanius ludovicianus ludovicianus*) and a northeastern migrant subspecies (*Lanius ludovicianus migrans*) over roughly the same time period. I suggest that a clue as to how this might occur with such dispatch can be derived from a powerful argument presented by Darwin in his *Origin of the Species*. Therein, he devotes extensive discussion to the concept of artificial selection, the process used by breeders of animals and plants to produce the characters that they want in their domestic stock: wolfhounds and Chihuahuas, Manx and Siamese cats, Riesling and Cabernet, and so on. In doing so, he presents a compelling case that if there is existing variation in a character among members of a population, a selecting agent can significantly increase the expression of that character within the population in a matter of just a few generations. This can be done simply by allowing those members that have the desired character to breed with each other and preventing those that do not have it from breeding at all. In other words, by manipulating the fitness value of a character.

Darwin's point was mainly to make the case that somehow there was a way for expressions of characters to be passed from one generation to the next, a point that is now well understood thanks to Gregor Mendel, James Watson, Francis Crick, and many others. Darwin also wished to emphasize that the environment could serve in the role human breeders took in designing the phenotypes of various animals and plants. The point as it relates to our discussion, however, is that if there is a selecting agent that will choose one variant of a character over another, then evolution can occur very rapidly indeed.

I suggest that avian females can serve as that selecting agent. As discussed in Chapter 6, females exercise vast control over male characters of appearance and behavior due to the fact that male birds cannot force females to copulate. Considering this idea as it might relate to rapid change in the appearance of members of a migratory population as opposed to a resident one, I suggest that if those females in the migratory population that choose to breed with males having a different appearance from that of males in the nonmigratory population produce more offspring than those that choose to breed with males of "resident bird" appearance, it would not take many generations to produce measurable differences in appearance between the two populations.

Perhaps there is no better illustration of the potential power of sexual selection than spring departure date for long-distance migrants from their tropical wintering grounds. The genetic nature of departure timing was confirmed through an extraordinary series of experiments performed by Eberhard Gwinner and his associates at Germany's Max Planck Institute of Ornithology, in which they demonstrated that a migrant that breeds in northern Eurasia and winters in southern Africa (Willow Warbler) would commence spring migratory restlessness (normally signaling African departure) on very specific dates even when held in captivity throughout the year in the Radolfzell lab on the shores of Lake Konstanz in southern Germany. The genetic aspect of this behavior has been further verified through remarkable work on protandry by Jessica Deakin and her colleagues at the Advanced Facility for Avian Research, University of Western Ontario.

American Redstart males depart from their wintering grounds in Jamaica and arrive on their breeding grounds on average by a week or more earlier than females. This phenomenon of early male arrival, referred to as "protandry" in the literature (discussed in Chapter 4), is common to many species of long-distance migrants. Pete Marra, a former Smithsonian associate of mine, and others have argued that this behavior arises from male dominance of females in competition for critical wintering food resources, which brings about superior physiological condition allowing males to depart (and arrive) at an optimal (earlier) time than females. This theory assumes, of course, that genetically programmed timing for departure is the same for both sexes.

Dr. Deakin and her confederates sought to test this hypothesis with Black-throated Blue Warblers, another long-distance migrant demonstrating protandry. Males and females were captured during fall migration, held in captivity over winter, and photostimulated in the spring to induce *Zugun-*

ruhe (signaling departure intention). Although held under the exact same conditions as females, with food readily available, males initiated departure intention five to eight days earlier than females, a timing difference roughly equivalent to observed arrival differences between the sexes on their breeding grounds in the northeastern United States. Barbara Helm and her coworkers at Radolfzell found a similar pattern for stonechats in which males initiated spring *Zugunruhe* an average of 4.6 days before females.

These kinds of experiments are difficult to perform. They require superbly trained and funded "pure science" research groups with outstanding institutional logistical support, few of which exist in the world. Thus, we have only a few sets of experimental results from a small number of species on which to base our theories; yet, the data that we do have in this case demonstrate that protandry is not a result of differences in competitive ability between the sexes for critical resources during the wintering period. Rather it is due to differences in male and female genetic programs for timing of departure and breeding-ground arrival, at least for this species, which is a critically important finding.

This discovery, however, begs the question of *why* males should be programmed to arrive earlier on their breeding area. I argue in Chapter 6 that the reason for protandry programming is sexual selection. Females choose males as mates based, at least in part, on the "quality" of their territory in terms of potential nest sites and food provision. Thus males must arrive earlier than females to assure that they have staked out and defended against other males the highest quality territory that they can. Failure in that effort could result in zero fitness (no offspring) for them. Partial migration, in which adult males remain on the breeding ground whereas adult females and juveniles depart, may result from a similar dynamic.

Margaret Morse Nice first published parts of her classic *Studies in the Life History of the Song Sparrow* in 1933–34 in the German periodical *Journal für Ornithologie*, wherein she detailed movements and other behaviors of members of the different age and sex groups of the eponymous sparrow based on her eight-year study of a population located near the Ohio State University campus. This bird is a short-distance migrant in Ohio, which means that most migrate less than a thousand miles from their breeding area to their wintering areas in the southeastern United States (Georgia, South Carolina, etc.). Among her fascinating findings was the observation that most adult males remained on or near their breeding territories throughout the year, including during the winter months, whereas adult females and juveniles of both sexes migrated south in October.

This pattern of seasonal movement, called "partial" or "individual" migration, has since been found to be common among other short-distance migrants including the Dark-eyed Junco, European Robin, Field Sparrow, Eastern Bluebird, and many others. Sid Gauthreaux, in his seminal review of the topic of migration for the multivolume *Avian Biology* series, ascribes partial migration, as well as the tendency for males of longer distance migrants to winter closer to their breeding range than females, to male dominance; that is, males choose the optimal distance to migrate for winter survival forcing females to migrate farther than they would otherwise choose in the absence of males. This theory is supported by the observation that, often when males and females of several short-distance migrant species are observed at the same food source during the wintering period, males limit female access through direct attacks and chases.

There are, however, some problems with the dominance hypothesis as an explanation for partial migration. First, as described in Chapter 4, females often winter in different places or habitats than males, thereby avoiding competition altogether. Second, there are no good data on differential survival between the sexes for any short- or long-distance migrant of which I am aware, regardless of where they winter or what habitat they winter in. Third, there are no data allowing us to compare survivorship for males remaining on their breeding territory as compared with males that migrate south. In the absence of such data, it must be admitted that the key elements of the dominance theory, namely differential survivorship values for the sexes or migrants versus nonmigrants, are no more than assumptions.

Well, okay, but what other explanation is there? The answer, again, might be sexual selection. If male fitness is dependent upon female choice of males with the highest quality territory, then males should do whatever is possible to maintain control of such a territory, even at the cost of potentially reduced survivorship resulting from attempting to continue defense through the winter months despite limited access to food.

Sexual selection may also help us to understand how migrants could react much more quickly to climate change than has been predicted by modelers. The work by Gwinner and his associates has demonstrated that departure from the wintering ground on spring migration for long-distance migrants is under genetic control. Therefore, change in timing of such departure for these "calendar" migrants should require evolutionary time (eons?) in which to occur, despite the fact that the environment on the breeding ground favors much earlier arrival due to global warming *right now*. I suggest that if variation in departure timing already exists in a population of calendar migrants, female choice of earlier arriving males as mates could exert the kind of selection pressure necessary to cause quick change in arrival times. There is already some evidence of such rapid change from various studies of male passage dates along the migration route and breeding-ground arrival timing.

Balancing Selection

> There saw I Sisyphus in infinite moan,
> With both hands heaving up a massy stone,
> And on his tip-toes racking all his height,
> To wrest up to a mountain-top his freight;
> When prest to rest it there, his nerves quite spent,
> Down rush'd the deadly quarry, the event
> Of all his torture new to raise again;
> To which straight set his never-rested pain.
>
> —Homer, *The Odyssey*
> (George Chapman translation)

The challenge facing populations to achieve optimal fitness through the effects of natural selection has been likened to that of Sisyphus in that the environment is constantly changing, thereby shifting what is, in fact, optimal. This situation is true, of course, for all organisms living on earth, or presumably anywhere else. Nevertheless, one can readily understand how much more complicated the process becomes for a migratory species confronting several remarkably different environments over the course of the annual cycle. Thus the various adaptations shown by migrants must represent a balance between what is optimal in the different selective environments. I think that this aspect of migrant evolution is well illustrated by the problem confronting females in those species in which males are more brightly colored (about a third of North American migrants). Obviously, there are very good reasons for the female to be cryptic during the breeding season, when a large variety of predators is searching for nests with eggs and young. In addition, she need not worry about competition for food with males, not only because the male's fitness is tied to her well-being but also because food is superabundant. However, as discussed in Chapter 6, all bets are off once the young reach independence. Individuals must fend for themselves. When food is limiting, as it often is during the wintering period, this situation requires females to compete not only with other females but with males as well. As demonstrated in Chapter 4, such competition can include territoriality in which the displays used in defense are often highlighted by the male's coloring. In these instances, the female is at an evident disadvantage when competing head-to-head with males for the same resource. So here is an instance where balancing selection presumably comes into play. In Chapter 4, I mentioned several ways in which females attempt to reduce or avoid nonbreeding-season competition with males, such as wintering in different places or different habitats and so forth. But the fact that each of these choices comes with some fitness cost is perhaps best illustrated by the portion of the population of females that mimic male plumage to a greater or lesser degree (as shown by the Hooded Warbler in Figure 4.6). Clearly, the female must balance (in a fitness sense) the disadvantage of her cryptic plumage in competition with males during the wintering period against the advantage of reducing nest predation during the breeding season. This scenario is meant to be exemplary of the kinds of compromises all migrants must face using the process of balancing selection.

Reaction Norms

I have argued that sexual selection may constitute a powerful force for rapid evolution in migrant characters and behavior, which could help to explain some very curious observations from the field. But there is something going on with migrants that is even more curious than this: individuals often show behaviors presumed to be under genetic control that are expressed differently under different environmental circumstances. How can this be? The short answer is, we don't know. In fact, we are only beginning to identify under what circumstances this curious situation occurs.

Long before Watson and Crick identified the specific molecule that carried information from one generation to the next (in 1953), field biologists, at least from Darwin and Wallace on, were well aware that some behaviors appeared to be under genetic control. Scotsman Sir Arthur Landsborough Thomson, who famously summarized knowledge of the migration field in his *Problems of Bird-Migration* (1926), suggested as much with regard to timing of departure from the breeding ground. However, he noted, "If the racial custom is similarly inherited by all the birds [of any given migratory species], what is it that stimulates it to greater activity, or to different activity, as between one individual and another?" He further comments that "evidence is lacking as to whether, in cases like this [that is, where differences in departure timing are observed], any given individual behaves in the same way in successive years."

As we have noted above, Margaret Morse Nice, in her studies of Song Sparrows, confirmed that Thomson's suspicions that a bird that migrated one year might not migrate the next were correct. She found that nearly all females and juvenile males migrated south in October, whereas nearly half of adult males remained on their Ohio breeding-ground territories through the winter. Thus, males behave in one way (migratory) as juveniles and another (resident) as adults. She also found, based on her work with banded individuals, that some adult males migrated in some years but remained on their territories in others. She thought that weather might be a factor in the bird's decision because in some instances a bird would leave during a particularly severe period of low temperatures and snow, and then return some days or weeks later.

I have explained this tendency of adult males to remain on their breeding territories throughout the winter as a possible result of sexual selection, which implies, of course, that the behavior is genetically programmed. But then we have also discussed that timing of fall migration departure is genetically programmed. How is that an Ohio Song Sparrow male will follow one genetic program as a juvenile and another genetic program as a territorial adult? And how can that territorial adult male follow a program to remain on the territory through the winter under one set of environmental circumstances in one year and choose to leave under a different set of environmental circumstances in another year?

Once pointed out, of course, the differences under different environmental circumstances in expression of behaviors assumed to be under genetic control were quickly recognized in a large number of situations in many different types of organisms. These inexplicable phenomena are called "reaction norms"—different patterns of phenotypic expression of a single genotype under different environments. The term was assigned by Dutch ornithologist Arie van Noordwijk. Somehow, in ways that we do not yet understand, flexibility is built into the individual's genetic program.

But how is this possible? A gene is a gene is a gene. If you have the genes for blue eyes, you don't end up with a different color just because your mother read *The Green Eyes of Bast* (although admittedly a terrifying tale) while you were in the womb. Or do you? The point is that our current understanding of how evolution works provides no information on how a character or behavior under genetic control can be expressed one way under one set of circumstances and a different way under another. Be that as it may, it happens. And it happens a lot in migrants. Just to present one instance, Eberhard Gwinner and Barbara Helm found that among stonechats taken from the same nest some individuals left on migration, presumably for their African wintering quarters, whereas others did not, remaining on their breeding areas in Ireland through the winter.

Dr. Helm is a good friend and colleague of Dr. van Noordwijk's, and having had many discussions with Arie on the topic, and having done work on reaction norms herself as described above, she agreed to provide me with a depiction of the concept, shown in Figure 9.4.

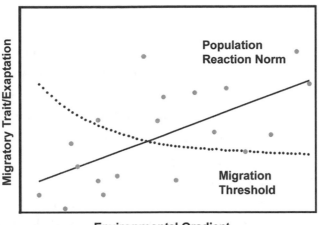

FIG. 9.4. Schematic illustration of a population reaction norm. Expression of the migratory trait for immediate departure on fall migration (*y axis*) varies over an environmental gradient that includes weather as well as individual status, for example, age, sex, or physiological condition (*x axis*). The slope indicates the mean departure response to environmental variation (i.e., the mean population reaction norm). Individual population members are shown by gray circles and scatter around the mean. Above a certain threshold value (*dotted line*), departure is initiated. BASED ON A CONCEPT DEVELOPED BY BARBARA HELM

Cultural Evolution

In addition to the modes suggested above for passing information from one generation to the next, there is at least one additional possibility, namely, direct communication. There is no doubt in my own mind that for humans this method has greatly surpassed all others, including DNA, at least since the invention of language. One need only consider the niche concept to confirm this fact. For all species other than humans (so far as we know), the niche—what an animal eats, where it lives, its courtship, its raising of young, and so forth (all the ways that the organism makes its way in the environment)—is fixed, genetically predetermined. For humans, we know this is not so. In fact, humans occupy tens of thousands of niches depending on the training of their parents and the society into which they were born. Thus a child can grow up to be a farmer, fisherman, rancher, warrior, priest, chef, accountant, doctor, lawyer, Indian chief, rich man, poor man, beggar man, thief, oligarch, and so forth, ad infinitum, depending on the kinds of information provided during their upbringing, their own inclinations and capabilities, and luck. But can important information be passed to offspring by their elders in other species? The answer to that is obvious. Of course it can. I have observed it myself, as have many of you. Consider feeder use by birds. When I was growing up in western New York, the species that would get food from a hanging feeder containing seed or suet was limited to a relatively small number: finches, chickadees, nuthatches, and so forth. I never saw a Pileated Woodpecker or Red-headed Woodpecker at a feeder, or even heard of one. Today, half a century later, such a sight is common in some places. What changed? First, more people put out food throughout the year. This change increases the probability that an individual bird will stumble upon the fact that feeders offer an easily harvested food source. Second, once this bird learned this fact, it showed its young how to locate and use it. I have seen this process in the works with Downy Woodpeckers. An adult downy shows up at the feeder, followed by a young one. The adult flies in confidently and begins to feed. The young one follows tentatively, searching at first for how to access the food, handle it, and consume it, keeping one eye on the adult all the time. Within a day or two, everyone in the family has the message and all fly in together, adults in their worn, year-old plumage and young in their fluffier juvenal plumage, all comfortably feeding together. House Sparrows go through similar training, except when the young are first led to the feeder by an adult, they simply sit on a perch and beg. The adults put up with this initially, picking up a seed and feeding it to the fluttering fledgling, but eventually refuse to feed them after a day or two of leading them to the source, forcing them to try it out for themselves. I suggest that it is as a result of this increased use of feeders year-round, and the training provided by adults, that we now see a vast new array of species at feeders—Indigo Buntings, Scarlet Tanagers, Rose-breasted Grosbeaks, Baltimore Orioles, Palm Warblers, Carolina Wrens, and so forth.

"Fine," you may say. "What's it got to do with migration?" In truth, I don't know, but there is extensive evidence that social cues can play a critical role in bird movements, presumably including migration for some species. Anyone who has stood outside under a low ceiling of clouds on a September night can tell you that nocturnal transients communicate. You can hear their soft twitters and chirps as they pass over, each unique to a given species. Does this involve intergenerational communication? And if so, is it important? Again, I don't know. I simply mention the possibility. I only know that it is a lot easier and safer to be told right from wrong than to learn it on your own.

CHAPTER 10

BIOGEOGRAPHY

There's no place like home.

—Dorothy, *The Wizard of Oz* (1939)

Biogeography is the study of the geographical distribution of plants and animals, or for us in particular, that of migratory birds. The fundamental question of biogeography is "why is any given species found where it is?" The answer for migratory birds that I will suggest in this chapter has three parts: (1) dispersal, (2) climate change, and (3) seasonality. As noted in Chapter 9, I have proposed that dispersal is the origin of a migratory habit. But as is clear from the parable of the sower, dispersal is a chancy process. "Good ground" is not always easy to come by. And that is where climate change comes in. Climate change creates new environments. If these new environments offer "good ground" for breeding, and they are aseasonal, then a simple expansion of the species' range occurs. However, if the "good ground" is seasonal, then occupancy occurs only during the breeding period, followed by return to point of origin once breeding is completed. Or at least that is what I propose.

Migrants, of course, are not seeds. They have wings, and if they don't find good ground, they can come back home. Even if they do find good ground, they can always come home if it goes bad, that is, is only seasonally good. If this theory is correct, then climate change is the chief determinant of migrant biogeography, creating new seasonal environments that dispersing birds can readily exploit, or completely obliterating previously existing environments. In recent times, that is, within the past few million

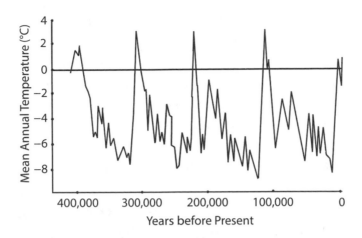

FIG. 10.1. World mean annual temperature variation over the past four hundred thousand years in degrees centigrade. Temperature peaks are interglacial periods, lasting about ten thousand years, followed by glacial periods, lasting roughly one hundred thousand years. We are currently in an interglacial period. BASED ON RETTIG 2019

years, climate has undergone massive cyclic changes with profound effects on migratory birds.

Current understanding postulates five major ice ages in the history of the planet, the most recent of which began about 2.4 million years ago. The current, or Pleistocene, ice age—the one in which we are living at present—is characterized by alternating sequences of cooling (glacial periods) and warming (interglacials) (Figure 10.1). Best estimates are that there have been at least ten such alterations between ice ages and interglacials over the past million years. No one knows the exact cause for sure, but perhaps the most accepted theory at present attributes the ultimate cause to two astronomical events: variation in the earth's orbit and the angle of tilt in the earth's axis relative to the sun. The idea is that some combination of these has produced the observed cycles, although why these produced an ice age after a twenty-million-year period of gradual cooling is unknown, or why the cycles have slowed from every forty thousand years or so two million years ago to the current one hundred thousand years is unknown.

Pretty much everyone agrees on these basic points. It's when we talk about the current interglacial that things begin to get contentious. Let's start by what we can all agree on. First, the last glacial maximum occurred about seventeen thousand years ago. Since then, global warming has been underway with resulting dramatic shrinking of the ice caps. It is this shrinkage, the result of massive change in climate in North America, with half the continent going from being covered by ice to half the continent being covered by forest, that has fueled invasion of over one hundred species of migrants, as discussed in Chapter 9.

Now we get to the part where people start fighting. It's hard to believe in today's superheated political climate, but a scant thirty-five years ago nobody but a few egghead scientists had ever considered the possibility of rapid, human-caused climate change (known at that time by the shorthand phrase the "greenhouse effect"). The trigger point for when the topic took off as a major public policy issue can be precisely dated. It was June 24, 1988, when James Hansen, then director of NASA's Institute of Space Studies, presented testimony before the Senate Energy and Natural Resources committee. During that testimony, Dr. Hansen stated that he was ninety-nine percent certain that the warming trends recorded by scientists worldwide over the past century were not natural variations but resulted from carbon dioxide buildup in the atmosphere from human use of fossil fuels.

At first, the US Congress was impressed, as shown by this quote from the committee chair calling for immediate action: "As I read it, the scientific evidence is compelling: the global climate is changing as the earth's atmosphere gets warmer. Now, the Congress must begin to consider how we are going to slow or halt that warming trend and how we are going to cope with the changes that may already be inevitable."

And indeed, they did take immediate actions, including appropriation of funds to stimulate research on the subject.

Meanwhile, the backlash to Dr. Hansen's testimony was not long in coming. Initially, it came mostly from other scientists, mainly meteorologists, who contended that the climate data did not demonstrate conclusively that rising global temperatures represented a true trend. They argued that change was cyclical and that the cycles could be decades or even centuries in duration. In the late 1980s, this argument seemed at least possible, as acknowledged by Dr. Hansen himself by leaving one percent of wiggle room. It is interesting to look back on this controversy in light of how the issue has developed. Normally, in science, when you have two contradictory hypothetical explanations for a given phenomenon, you proceed to test the two. And that is what happened initially in this case. There were basically two camps of scientists: one thought that recent warming had been significantly affected by human activity and one thought that it was a result of unknown natural causes and that humans had little effect. And so scientists went out to test these hypotheses. But they were confronted by some procedural problems. One was a lack of good long-term, worldwide historical climate data sets. In Hansen's testimony, he stated that 130 years of data showed that the rise in temperature was a trend, not a cycle. But 130 years of what data? How many countries have reliable temperature data for 130 years? And what if the cycle is eight hundred years or a thousand years? Also, back in 1988, there were very few super computers available to academic research scientists. Which means that we did not have the computer power to run the extraordinarily complex models required to perform accurate tests of the competing hypotheses. Also, back then, we really didn't have the knowledge to understand what questions to ask.

My first professional experience with the climate change issue came in 1990 when I was a new research scientist at the Smithsonian Institution. Part of the congressional appropriation that I mentioned earlier was made available to Smithsonian units in the form of competitive grants. Colleagues and I put together proposals to try to get some of this "global change" money—and we did get some. But we didn't really know what we were doing. It was too new. The money ended up going mostly to new research positions at the Smithsonian—not necessarily a waste, but not really advancing knowledge on climate change either.

Another major problem was that many scientists forgot that they were scientists and started making incautious statements regarding the future.

This behavior is natural. It is human nature to enjoy attention, especially among scientists, most of whom are not used to being on *Oprah*. They like the limelight, and they forget that their job is to formulate and test hypotheses, publish the results, and create new hypotheses for testing. When they get into the arena of forecasting, they risk their objectivity. As soon as you say, "I think this is going to happen," you have a vested interest in being correct. As a scientist, that's a very big problem because when you look for something, you usually find it. Biases have a very bad habit of creeping into analysis and conclusions. Hansen, and many others, made a lot of such statements early in the game, which were misguided at best.

For instance, Hansen predicted that the paper birch would disappear from the northern United States in a matter of decades. Now, nobody likes to hear that. The paper birch is a signature of New England fall foliage. It's the state tree of New Hampshire. Such pronouncements grab the public's attention, which is what they are supposed to do. The problem is, what the heck does James Hansen, admittedly a world-renowned climatologist, know about paper birch ecology and the possible effects on its distribution of global warming? The obvious answer to that is very little. In fact, evidently, less than many members of the general public, who would know from personal observation and experience that, even though the natural range of the paper birch coincides with taiga or boreal forest, they are grown as ornamentals across large portions of the country far outside their natural range, including many places that are very warm for long parts of the year.

Scientists did not have the information to make accurate forecasts on how rising temperatures might affect the world's climate, weather, or ecosystems back in 1988, but that did not stop at least some of them from making such forecasts. In their defense, many whom I knew felt that the threat was of such magnitude as to require dire warnings—even if they turned out to be inaccurate. The outcome in terms of the framing of the debate, however, was not good. It resulted in a loss of confidence by the public in the science of climate change in general, and a loss of focus in the public discussion regarding what was actually known.

By 1995, the problems of testing the cycle hypothesis had been largely resolved. Hundreds of scientists working independently on various aspects of the question had published peer-reviewed papers documenting that global warming is a trend, not a cycle. In addition, by early 2000, models had been developed that made a strong causal link between carbon emissions and global temperature rise.

However, even this fundamental point is not settled in the public's mind. What the public knows is that there are two groups of experts. One claims that the end of the world is coming because we are heating our houses and driving cars. The other group argues that the end of the world is not coming and evidence of global warming is bogus—and even if it isn't bogus, it won't be that bad; and even if it is that bad, there isn't anything we can do about it. Presently, there is scientific consensus that significant anthropogenic-carbon-emission-related temperature change is a cause of recent climate change, although there are many researchers who have expressed valid questions regarding the methodology and the accuracy of the conclusions for a number of the relevant studies (questions cogently expressed in Dr. Donald Rapp's book, *The Climate Debate*).

Personally, I don't know.

Several scientists have suggested that warming temperatures, per se, could have a direct effect on bird species distribution. In fact, I recently read a National Science Foundation proposal designed to test this idea as it might pertain to changing breeding distribution of Black-throated Blue and Hooded warblers in the mountains of Georgia, in which the researchers plan to assess the viability of eggs from nests of these two migrants under different conditions of temperature and humidity (comparable to the kinds of changes that might be expected from global warming in the region).

Many similar kinds of experiments likely are underway, but I doubt that direct effects of temperature will explain the rapid observed changes in distribution for most of the animal and plant species in which it is currently happening, let alone birds. Indeed, it seems highly unlikely to me that temperature has any direct effect on bird distribution. Birds are among the most physiologically competent organisms on earth in terms of thermoregulation. Thus, if there is a relationship, it must be indirect. That is, change in climate is causing something to change that is important for a broad spectrum of bird species—perhaps insect hatch dates or frost-free days or fruiting periods or who knows what?

I don't.

And what is the cause of recent climate change, anyway? As discussed above, there has been a great deal of "heated" discussion on this issue, with the scientific community largely concluding that the cause is human consumption of fossil fuels and the resultant release of large amounts of greenhouse gases, such as carbon dioxide and methane, into the atmosphere, where they remain, serving as an insulator for the earth, and resulting in global warming and climate change. But what would have happened had humans not been here, as was the case for all of the previous interglacial warming periods over the past two million years? Why did they occur? How fast did they occur? We are only beginning to get the answers to these questions, but even the initial data are enough to raise questions concerning the role of humans in the process. Humans were not around during the previous nineteen instances of warming periods. But fire, of course, was. It may be that massive, unchecked burning of forests, prairies, and tundra could explain the final stages of warming during previous episodes.

Of course, birds care about none of this. Their focus is on maximizing their genetic contribution to the next generation. Our Welder work on shifts in breeding distribution demonstrates that birds are not waiting for us to figure out whether or not change is occurring, and, if so, what they should do about it. Members of at least eighty species of birds are rapidly occupying new breeding ranges—some as permanent residents and some as migrants.

So? What's the big deal about a little range change? To comprehend this, you need to understand that, in biological terms, a species' range has long been considered to be a genetic trait, like size, color, or habitat use. It is not simply where an organism is found but a characteristic of a species that is also a critically important theoretical construct. Recall that Ernst Mayr defined the biological species as "groups of actually or potentially interbreeding natural populations reproductively isolated from all other such groups." The key point here for us is "reproductively isolated." Organisms are not supposed to stray outside their ranges, certainly not in terms of spreading their seed. Indeed, the range has been considered the keystone of the biological species concept as formulated by Dr. Mayr, and it is this requirement that individuals not move beyond their range that has always constituted the chief weakness of the theory. According to the biological species concept, dispersing individuals from a given population must stop dispersing at the edge of their range. Supposedly, they are greatly assisted in this discipline by obstacles, such as mountains, rivers, oceans, or simply space. Indeed, according to population theory, genes of a single individual from a neighboring population in a given generation are sufficient to maintain two separate populations as members of the same species (that is, the populations are *not* reproductively isolated from each other).

Many biogeographers have known that this requirement simply was not true for many closely related species for a long time, in fact, from well before when Mayr formulated his idea. The evidence was perfectly obvious. One of the regions most clearly demonstrating the fallacy of the argument occurs along the border between the subtropics and temperate regions in South Texas, where dozens of species of plants and animals (including many birds) show one species on the temperate side of the border and another on the subtropical side, despite complete absence of a physical barrier of any sort and no evidence of one over geologic time covering the assumed age of the species involved. Examples among birds of such species pairs include the Black-crested Titmouse (subtropical) and Tufted Titmouse (temperate), Tropical Parula (tropical) and Northern Parula (temperate migrant to the neotropics), and Great-tailed Grackle (subtropical) and Boat-tailed Grackle (temperate), among many others.

So what is it that maintains these sister populations as separate species if not prohibition of interbreeding? I suggest that sexual selection could play a significant role. Due to the significant ecological differences that exist between the regions, differing adaptations might be expected to evolve quickly given a selecting agent. Females could serve as that agent, with those females that chose to mate with males having the adaptations appropriate to their environment producing more offspring that contribute to the next generation than those choosing the "wrong" males as mates. Just a suggestion. Actually, it doesn't matter much whether I am correct or not. The facts on the ground are clear. It's up to the theoreticians to explain why.

How does all this relate to migrant biogeography? The answer is that the movement capabilities of dispersing individuals is at the heart of my migration theory. If dispersers limited their movements to within the existing range of their species, as required by Mayr's theory of biological species, there would be no range change or migration. For my theory to be correct, individuals must be capable of moving long distances from where they were born. I have presented several different kinds of evidence to document that all birds, whether resident or migrant, possess this ability. In this regard, the Texas range-change data are especially telling. As soon as climate changed the nature of the habitats outside their range, to the north and east, over eighty species of birds expanded their breeding ranges to include this newly created environment in a period of less than thirty years. How could this happen? Obviously, they couldn't "know" about the change. So the only way it could happen is if the "seeds" (that is, dispersing individuals) put out by the great "sower" (random dispersal) began to find "good ground" where none had existed before.

Thus findings from our work at Welder are very exciting. For me, they show exactly how invasion of environments newly created by climate change must work—just as it must have occurred eight thousand years ago as the glaciers retreated and forests made their northward advance. What is so remarkable is that responses by animal and plant species take place in real time (years), not geologic time (millennia). An individual researcher can actually record and measure changes in distribution as they happen.

Eight thousand years ago mammoths roamed across periglacial steppe within ten miles of where I sit today in the midst of second-growth forest of maple, hemlock, beech, and oak—forest filled with forty species of long-distance migrants from May through August. I suggest that dispersal by birds into the rapidly changing environments resulting from global warming over the past few thousand years created these diverse communities of long-distance migrants whose parent populations likely are derived from the tropics (where many still have breeding populations). Thus, in my opinion, climate change associated with the recent ice age is the principal shaper of the diversity and distribution of our current migratory populations.

I am not the first to suggest such a relationship. The diversity of songbird migrants in the Western Hemisphere has long been a subject of keen interest for biogeographers, and ice ages have often been accorded a critical role, perhaps most famously in Dr. Robert Mengel's classic 1964 paper in the *Living Bird* on ice age effects on speciation in the avian taxonomic group of wood warblers. His opening paragraph states the matter well:

> Zoologists today are virtually agreed (Mayr 1963:480, 513) that species formation in warm-blooded vertebrates is a process requiring geographical isolation of stocks ("allopatry") long enough for development of morphological, physiological, or psychological isolating mechanisms adequate to insure continued genetic separation in case of renewed coexistence, or "sympatry." How exquisitely complicated, then, must be the history of geographical isolations and distributional changes resulting in the many complex arrays of closely related genera and species of birds found in many parts of the world.

Mengel goes on to lay out in convincing detail how species groups of warblers seemingly fit the potential isolating mechanisms provided by repeated glaciations over the past two million years. The basic pattern, illustrated by the Black-throated Green Warbler group, is one in which a presumably ancestral bird, the Black-throated Green in this case, has a broad breeding distribution in coniferous and mixed forest across the continent prior to the initial glaciation (2.4 million years ago). This glaciation covers large portions of the forested region of the continent with ice, resulting in the creation of pockets of forest habitat in what is now the southeastern and southwestern United States. Due to the nature of the geography of these regions, a single pocket of suitable forest was formed in the southeast, whereas the mountainous southwest resulted in the establishment of several mutually isolated pockets. The result was the evolution of three or four species of western warblers, derived from and closely related to the single, ancestral species of eastern warbler. The maps depicting pockets fit during the glacial maxima, and the assumed isolated distributions of the members of the group were extraordinary (Figure 10.2), but the clincher was the fact that the distributions for more than one warbler group fit the picture.

The presentation was convincing, especially for those of us with limited knowledge of the timing of Pleistocene glacial and interglacial periods and the distribution of habitats during the various glacial advances and retreats, which, initially at least, was just about everybody. Paleogeographers would likely have seen problems with the paper, even back at the time when it was written. But if any read it, they likely would have figured that the bird boys were just off on one of their harmless, fanciful flights, wasting taxpayer money on their research. "They will eventually get it right in a half century or so, or not," they might have thought.

I think that I first read the paper in the early 1980s when I was trying to find everything that I could on neotropical migrants for my book on the topic (*Nearctic Avian Migrants in the Neotropics*). What struck me at the time was that, although Mengel recognized that wood warblers were of tropical origin, his maps and word models paid no attention to the winter range, which actually included part of the breeding range for some of the species treated, like Grace's Warbler. Speciation was considered to be an entirely breeding-ground affair. This idea seemed completely wrong to me because my tropical work had convinced me that the critical behaviors, structures, and physiological adaptations for these birds evolved in the tropics. The birds moved north to capitalize on seasonally superabundant foods that required little in the way of specialization to harvest. Accordingly, I developed my own models to explain Mengel's findings that took into

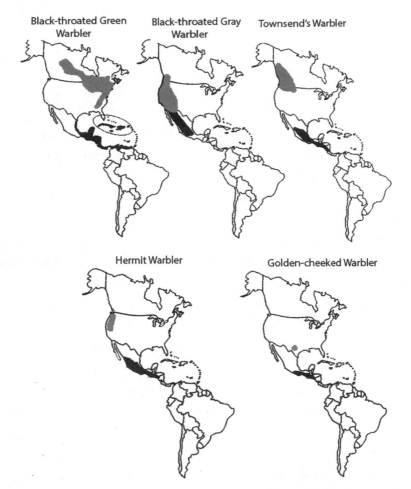

FIG. 10.2. Distribution maps for species in the "Black-throated Green" group. Breeding = gray; winter = black.

account winter distribution. I think that I was right in pointing out importance of the wintering ground in migrant speciation, but I was completely wrong to accept Mengel's Pleistocene habitat re-creations.

It was two former students of mine, Bob Zink and John Klicka (they've gone a long way beyond anything I ever taught them), who blew Mengel's model completely out of the water in a justly famous paper published in

Science in 1997 entitled "The Importance of Recent Ice Ages in Speciation: A Failed Paradigm." In this paper, they demonstrated that, assuming that mitochondrial DNA evolves at a rate of two percent per million years, genetic distances between the various species involved in Mengel's treatment were far too great to have occurred during the Pleistocene.

As I have mentioned, I do not accept that population divergence is a random event. I believe that sexual selection can play an important role, and that it can occur far more rapidly than has been assumed. Nevertheless, whether or not their timing calculations were correct, Zink and Klicka's work was not the only problem with Mengel's hypothesis. Mengel based his work on an assumed number of four glacial periods and accompanying interglacials during the past two million years, and his habitat reconstructions were, as he admitted in the paper, somewhat hypothetical. We now know that there have been twenty alternating periods of glacials and interglacials, and that Mengel's data on the character, number, and timing of habitat remnants during glacial maxima were not simply hypothetical but wrong.

This point was made in another important paper by Zink (and Aubrey Gardiner, who is not a former student of mine) published in 2017, which showed that "70% of currently long-distance migrant species lacked suitable breeding habitat in North America at the Last Glacial Maximum (LGM)." In other words, migrants did not have safe little islands of temperate zone breeding habitat where they could go about their lives, gradually differentiating from their brethren in similar, isolated pockets. Instead, as Zink and Gardiner suggest, members of these species likely resumed breeding in the winter ranges from which they were ultimately derived in the first place. This hypothesis makes sense to me, although I disagree with them regarding how these tropical birds resumed their temperate migrations during interglacials. They suggest, following the Migratory Syndrome hypothesis, that migration is a simple genetic switch, turned off and on based on the selective environment. As I have argued elsewhere, I do not believe any such switch is required. All that is needed is dispersal and a new environment.

But let's leave that for the moment. So what accounts for those fascinating warbler groups pointed out by Mengel? I think that he was right in emphasizing the importance of western mountains but wrong in their location. I think that these speciation events likely resulted from the different populations of members of these warbler groups becoming isolated in islands of appropriate habitat on the mountains of tropical western Mexico.

The Migratory Syndrome

> But he hasn't got anything on!
>
> —Hans Christian Andersen, *The Emperor's New Clothes*

In the summer of 2020, I was assisting my friend Jim Berry, former executive director of the Roger Tory Peterson Institute, with some bird banding. Jim had brought along a recently published book sent to him for review, the *Peterson Guide to Bird Behavior*, by eminent tropical ecologist and author John Kricher. Jim advised that I should take a close look at the section on migration, which I did. Therein, Dr. Kricher espouses the Migratory Syndrome for the origin of migration in the Western Hemisphere as proposed in the 2014 paper in the Proceedings of the National Academy of Sciences by Benjamin Winger, Keith Barker, and Richard Ree. This observation surprised me. I was well aware of the fact that in the six short years since that article's publication, it had become the predominant paradigm for the evolution of migration, but to find that Dr. Kricher was a believer came as a shock. Like me, he cut his metaphorical teeth studying the ecology of migratory birds in the tropics, predominantly Belize, I think. But there is no ecology whatsoever in Winger et al.'s version of the Migratory Syndrome.

The Migratory Syndrome hypothesis was developed by the Australian entomologist V. Alistair Drake and colleagues in a book published in 1995. Therein, the syndrome is described as "a suite of traits enabling migratory activity." The concept was developed for application to insect migration, but it has been widely applied to avian migration as well, originally by Hugh Dingle, who published in 1996.

The theory assumes that migration results from an ancient gene or attribute (several genes) that controls a highly integrated bundle of adaptive traits (*Zugdisposition, Zugstimmung*, navigation, orientation, etc.) generally assumed to have evolved over eons of selective pressure, likely imposed by gradual change in the climate from warm and more or less aseasonal to cooler and seasonal. In some versions, the gene or bundle is considered to be present in all or nearly all birds due to its possible origin in dinosaur progenitors or even earlier. This variant of the theory often includes the idea of a genetic "switch" that can be turned on and off by environmental conditions and pressures. In other forms, the bundle has evolved in several avian lineages at different times and places but also many millions of years in the past.

The beauty of the Migratory Syndrome from the perspective of molecular genetics is that, if it exists, you can make some very comforting assumptions regarding the validity of a phylogenetic tree in tracing its presence or absence in lineages back through millions of years of evolutionary history.

This approach was taken in the Winger et al. article. They propose that the first representative of the large group of birds that includes modern wood warblers, tanagers, buntings, and sparrows (Emberizoidea) arrived in the New World via dispersal and colonization from east Asia across Beringia (the huge swath of land joining the continents at the time) into North America roughly fourteen million years ago (Figure 10.3).

They argue that migrants in this group evolved from this proto-sparrow, which was originally a resident occupying northern climes when these higher latitudes were warmer than at present and perhaps aseasonal. Weather in these regions then deteriorated gradually during the winter period, pushing birds south seasonally, which, over millions of years, favored the evolution of migratory traits, such as prewinter fattening and departure timing. This process eventually resulted in the evolution of long-distance migration to the New World tropics for some populations (Northern Home Theory echoes), where some wintering members remained to breed ("migratory drop-off"), followed by subsequent diversification. The

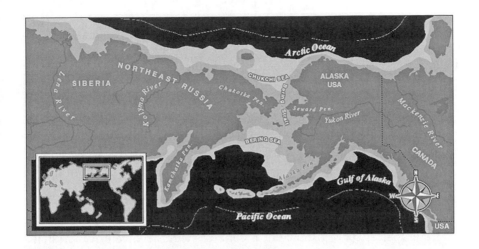

FIG. 10.3. Beringia: water = black; land = various grays with darker grays representing higher elevations above sea level. BASED ON ONLINE MAP FROM THE NATIONAL PARK SERVICE

large tanager family, Thraupidae (almost four hundred species, nearly all of which are tropical residents), is considered the premier example of this process in action. In summary Winger et al. state, "*We found that long-distance migration primarily evolved through evolutionary shifts of geographic range south for the winter out of North America, as opposed to north for the summer.*"

This statement is unique in my experience in reading thousands of scientific papers. Its singularity derives from the absolute certainty with which the results of their investigation are presented. To paraphrase, they simply assert "this is how migration evolved." In scientific terms, certainty means a zero probability of being wrong, which simply does not occur. Scientists know that there is no one hundred percent certainty. Newton was wrong. Einstein was wrong. Darwin was wrong. The universe is a complex place, and we have only begun to try to understand it. So Winger et al. are wrong, but how wrong? That depends on the known validity of their assumptions to the degree that we can figure them out. They believe that the assumptions that they admit to actually are rules, so if the rules are correct, and the analysis is correct, then their conclusions are as correct as two plus two equals four.

The problem with their conclusions is that they are based on literally hundreds of assumptions, almost none of which are stated, let alone defended. These can be lumped into five major categories: (1) origin of the emberizoid lineage in the Western Hemisphere, (2) timing of speciation events, (3) genetic distance required for creation of sister species from a parent species, (4) existence of a genetic switch evolved over millennia controlling migration, and (5) evolutionary changes in ecology and population dynamics required for annual movement between completely different ecosystems. Below, I consider (briefly—full exposition would require volumes) the bases and validity for each of these critical groups of assumptions.

1. *First New World appearance of the emberizoid lineage occurred in the north temperate region of the Western Hemisphere.* This assumption is based on phylogenetic analysis of the 832 species comprising the New World songbird group that includes tanagers, warblers, buntings, and sparrows reported in a separate paper (Barker et al. 2015). While it is true that such an analysis can give you a rough idea of when the first member of the group appeared (fourteen million years or so ago, give or take a million or two) and when diversification might have taken place, it cannot tell you *where* in the world that appearance occurred. It could have been anywhere, and, given the overwhelmingly tropical nature of the group, it was most likely in the neotropics. Thus, certainty that the group derived from a temperate sparrow by dispersal across Beringia is not justified without actual data or a large number of additional assumptions.

2. *Timing of speciation events.* The models and decision matrices used by Winger et al. to build their understanding for how, where, and when diversification occurred require a precise, standard timing, as explained in Appendix 2. Whether such timing exists is highly unlikely. Based on what

we see in the real world, speciation timing can change with varying degrees of environmental change, such as might occur during Andean orogeny or repeated Pleistocene glaciations.

3. *Standard genetic distance required for creation of sister species from a parent species.* As explained in Appendix 2, a standard genetic distance must be assumed in order to build the decision matrix that leads to the main conclusions of the article regarding origin of migration. However, we know that there is no standard genetic distance between sister species: some sister species show significant difference while others show very little.

4. *Existence of a genetic switch evolved over millennia controlling migration.* The hypothesis of temperate origin for a primarily tropical group of birds requires that these birds evolve a migratory habit. Winger et al. propose that the suite of characters necessary for successful migration evolved gradually over millions of years of deteriorating winter weather in the temperate regions, and that this suite was under the control of a genetic switch that could be turned on and off as environmental conditions required. As discussed elsewhere at length in this volume, no such switch has been found, nor is it likely that it exists.

5. *No evolutionary changes in ecology or population dynamics are necessary for temperate zone residents to invade tropical bird communities either seasonally or as new permanent residents.* It is with this huge bundle of assumptions that the Migratory Syndrome parts company with the Northern Home Theory (and occasioned my dismay with its ready acceptance by colleagues who have spent their careers investigating tropical ecology).

The ecological problem was neatly summarized by Reginald Moreau, a British field biologist who devoted his career to understanding the European-African bird migration system. He recognized that if migration originated with European species flying south to Africa to escape bad weather, this situation created an enigma ("Moreau's paradox"). How could birds evolved in temperate regions enter the complex ecological communities of the tropics and remain there for months? He suggested two possibilities: (1) *empty niches*, that is, no tropical species fills the niches the northern birds will occupy; and (2) *superabundant food*, that is, there is more food available than the resident birds can use. Robert MacArthur recognized this problem as well, as discussed at length elsewhere in this volume, and devised three explanatory devices: superabundant food, the "ecological counterpart," and the "fugitive species." Winger et al. take no pains to explain how this paradox could be resolved. They simply say that it was— many times over millions of years. In fact, they say that these northern invaders not only stayed in tropical communities as long as they wanted but established resident breeding populations as well ("migratory drop-off").

In summary, the actual data contained in Winger et al.'s Migratory Syndrome, as they pertain to migrants, present a situation similar to Russia's war intentions as described by Winston Churchill: "A riddle wrapped in a mystery inside an enigma." Phylogenetics most certainly will help us to understand how migration works, but not without grounding in the real world of where and how migrants live and have lived. How emberizoids first appeared in the New World tropics is an interesting question, but there is no ecological or genetic evidence to support the hypothesis that their entry involved a "migratory switch."

CHAPTER 11

CONSERVATION

The death of one man: that is a catastrophe. One hundred
thousand deaths: that is a statistic!

—Kurt Tucholsky, "Französischer Witz"

Conservation is not science. This point can hardly be better
illustrated than by the history of DDT use in the United States. By the early
1940s, this pesticide was in wide use for the control of mosquitoes and
other noxious insects. By the mid-1940s, thanks to the work of Clarence
Cottam and other field biologists, the devastating effects of this compound
on aquatic environments was already well understood, and by the mid-
1950s entire populations of birds high in the food chain (such as falcons,
hawks, eagles, and pelicans) were known to be disappearing throughout
the country as a direct result of its use. Data documenting effects of DDT
on the environment were quite clear, but no conservation action in terms
of attempts to control its use was taken because in the public mind DDT
was a miracle compound whose benefits to society in terms of increased
agricultural yields and decreased disease spread were unquestionable. It
was not until publication of Rachel Carson's *Silent Spring* in 1962 wrested
domination of the public imagination away from those promulgating the
"wonders of modern scientific pest control" to those warning of "the deadly effects of industrial poisons on our lives" that public action was taken. By
1973, DDT use in the country was outlawed, and we had the Clean Air Act,

the Clean Water Act, and the Endangered Species Act. These outcomes were not a consequence of scientific investigation, rather they resulted from who controlled the narrative.

This story illustrates an important concept. Conservation is public policy as it relates to the environment; it is as subject to the beliefs, whims, and fears of the polity as any other aspect of governance. The part played by scientific truth is highly variable. Whoever controls the story at a given point in time controls actions taken. Science, defined as a search for truth through the accumulation of facts and testing of hypotheses, plays an equivocal role in conservation, which, by its nature, is value-driven, whereas science (supposedly) is not. Once values and beliefs become associated with the process, objectivity is lost. Thus, in a sense, "conservation science" is an oxymoron. As has been found on many occasions, when facts do not support specific conservation values or initiatives, they are ignored as quickly by conservationists as when the reverse is true for anticonservationists.

As an example, in May 2018, I was listening to an NPR show (*Living on Earth*) dealing with the Trump administration's gutting of the enforcement provisions of the Migratory Bird Treaty Act. The act became law a century ago and played a major role in restoring populations of many species decimated by hunting, as well as putting in place provisions for penalties for any future actions that threatened migrants, which now includes such things as oil spills, waterway pollution, and similar industry actions. The host questioned his guest, a research associate of the Cornell Laboratory of Ornithology, regarding what might constitute major threats to migrant populations. The guest listed climate change and cats among others. As I will discuss below, whether or not these issues pose threats to migrant populations is a matter of conjecture, not fact. Such use of hypotheses as proven for the purpose of supporting a particular viewpoint is an all-too-common phenomenon in conservation "science."

The purpose of this chapter is to examine the various factors responsible for bird deaths in light of whether these factors have been, are, or could be threats to the continued existence of one or more species of migratory birds (based on a combination of evidence and my opinion) and to consider what best can be done by societies wishing to conserve their migratory bird populations. Each major factor is treated within a section of its own. Inevitably, I suppose, I occasionally deviate from migrants as the conservation focus to humans, particularly where migrants simply share with humans the fate of potential victims (as in climate change) or are considered as the principal agents of human problems (as in pandemic annihilation).

Disease

> When this terrible disease reaches this hemisphere, I hope
> you are the first to die from it.
> —"Elizabeth," a primary care physician from New Jersey,
> in an email to me

Pestilence, leader of the four horsemen of the apocalypse, has been the scourge of humanity for millennia, killing more people than all other causes combined by a wide margin across the arc of our existence. The black death, small pox, yellow fever, cholera, dysentery, malaria, and their like have repeatedly decimated societies, while whooping cough, diphtheria, influenza, pneumonia, polio, tetanus, rheumatic fever, and so forth made childhood a terrifying minefield. The twentieth century saw the conquering of most of these horrors, but their ghosts linger. After all, the millions killed around the world by Spanish flu in 1918 is still within living memory. My own father had Spanish flu as an eight-year-old. Thus, it is not surprising that infectious disease remains high on the list of our ineffable fears, and that each new manifestation, such as Ebola, Zika virus, swine flu, and COVID-19, sends tremors across the planet.

Our knowledge of disease has developed at a remarkable pace, but, unfortunately, our understanding of causes (etiology) has lagged, perhaps because of the need to know not only the ecology of the agent but also the various intermediaries between agent and human sufferer, which can include a bewildering array of completely different kinds of organisms. And this is where migratory birds enter the picture.

WEST NILE VIRUS

I love the smell of malathion in the morning.

Caption of cartoon by R. J. Matson in the *New York Observer*

On August 12, 1999, a sixty-year-old man came to New York's Flushing Hospital emergency room complaining of fever, weakness, and nausea. By the August 23, five additional patients had come in with similar symptoms; two of them died. A remarkable team of New York public health physicians, led by Deborah Asnis, suspected a viral encephalitic and contacted the US Centers for Disease Control and Prevention (CDC). By the 3rd of September, CDC immunologists had confirmed the diagnosis, identifying an agent related to Saint Louis encephalitis virus. Public health officials recommended immediate intensive spraying of the area with insecticide, mosquitoes being the known vector. Rudy Giuliani gave approval, prompting an infamous cartoon by R. J. Matson in the *New York Observer* depicting the New York mayor in the garb and posture of Colonel Kilgore (from *Apocalypse Now*) over the caption, "I love the smell of malathion in the morning!" The action likely prevented a major epidemic in one of the most densely populated regions of the world.

Contemporaneous with the human outbreak, there was a massive die-off (thousands) of birds, mostly crows, in the same area of northeastern New York City that was home to most infected patients. Brilliant work by a team of virologists identified the cause for both the birds and people as West Nile virus, previously unknown in the Western Hemisphere.

The CDC convened an emergency conference at their Fort Collins, Colorado, facility within two months of the outbreak, inviting the leading experts in public health, infectious diseases, immunology, and virology to discuss what the likely scenario for West Nile virus in the New World might be, and what to do about it. I was invited as a person knowledgeable regarding migratory bird biology because migrants were suspected of being both the introductory and amplifying hosts, thus serving as the viral pool that had brought the pathogen to New York and from which mosquitoes could transfer it to humans.

The conveners asked me to consider two questions in my address to the conferees. First, assuming that birds were the likely introductory hosts, as had been indicated in studies of West Nile outbreaks in the Old World, what modes of entry to the continent might I suggest? Second, how far and fast was the virus likely to spread across the hemisphere?

I knew nothing about West Nile or any other virus when the CDC contacted me, but there was a significant literature on the topic, which formed the basis for my presentation. The pathogen was first identified from the blood of a woman from the West Nile region of Uganda in 1937. Since then, it had become one of the most widespread viruses in the Old World. Migratory birds had long been suspected as the principal introductory hosts for the following reasons: outbreaks in temperate regions generally occurred during late summer or early fall, coinciding with arrival of large concentrations of migratory birds; these outbreaks often arose among people living in or near wetlands where high concentrations of birds came into contact with large numbers of mosquitoes; the principal vectors from which the virus had been isolated were mainly ornithophilic (preferentially feeding on bird blood); antibodies for the virus had been found in the blood of many migratory bird species in Eurasia; migratory birds had been linked to transport of related viruses in the Western Hemisphere; West Nile had been isolated from some species of migrating birds; and viremia (active virus identified in migrant blood) sufficiently long-term to infect vector mosquitoes had been documented in several individuals of some migrant species. One added note of interest regarding European outbreaks was that, unlike in the New York eruption, large numbers of dead birds was not a characteristic, presumably because New World bird populations had developed no natural resistance to the virus in contrast with Old World birds.

Armed with this information, I suggested in my talk that if birds were indeed the means by which West Nile was brought to the New World, there were three ways I could see that it might have happened: (1) A small number of individuals of a few species, such as the Eurasian Widgeon, migrate regularly between breeding grounds in northern Europe and wintering grounds along the northeastern coast of North America. Perhaps one such bird became infected at its breeding (or hatching) site and brought the disease with it. (2) Cyclonic storms form in summer off the coast of West Africa, often traveling across the Atlantic and up the US east coast, occasionally bringing African birds trapped in the winds with them, perhaps including a bird with active virus (viremic) in its blood. (3) Thousands of birds are transported to and through the port of New York by air and sea. Those transported legally, whether for the pet trade or zoos, go through a period of quarantine on arrival, but the possibility that a mosquito might break that quarantine to feast on a viremic subject does not seem too far beyond the realm of probability.

As to the future of the pathogen, I said that if it could indeed be moved by viremic migrants, West Nile virus would be found throughout the Western Hemisphere within a matter of weeks, although its evident lethality for New World birds suggested that the virus might simply go nowhere and die out as mosquito populations went into dormancy with approaching cold weather.

A questioner asked me to give my opinion on the most likely mode of entry. In response, I suggested that an infected, imported bird seemed the most likely source, especially since the largest concentration of birds dying from the disease was on the Bronx Zoo premises, a few miles from the Queens neighborhood from which most human patients derived. Maybe an infected Egyptian Goose or some individual of a similar species added to the zoo's collection from the Middle East could have been the source (virologists had found that the New York strain of virus most closely matched Israeli material).

This response prompted a virulent tirade from a member of the audience, Dr. Tracey McNamara, then head veterinarian at the Bronx Zoo, the import of which was that such shameful speculation was ridiculous and irresponsible given that quarantine measures certainly would have prevented any such occurrence.

We were divided up into smaller discussion groups after the plenary session to consider what plans should be made for tracking the pathogen and dealing with the public health concerns emanating from its presumptive imminent spread across the hemisphere. For whatever reason, I was assigned to the virology group. During discussions, I noticed a curious thing. The person chosen to moderate, as well as most of the other participants, seemingly directed most of their attention to the observations and comments of a single, elderly gentleman. In conversation with a member after the group had broken up, I found out why. He was Berkeley distinguished professor emeritus W. C. Reeves, the acknowledged dean of US virologists and former PhD advisor for several of the leaders in the field.

The CDC asked me to prepare a paper subsequent to the meeting for publication in their journal *Emerging Infectious Diseases* (EID), summarizing my thoughts on the topic, which I agreed to do. However, Dr. McNamara had taught me an important lesson, namely that if you are going to speculate in a public forum about critically important issues, you better have a solid team that includes recognized experts in the relevant fields. Reviewing what I had learned at the conference, it seemed to me that my most critical needs were a virologist whose specialty was West Nile virus and an expert in waterfowl, which not only are common Middle Eastern exotics in zoo collections but also are among the most regular, if rare, interhemispheric migrants to the North American east coast.

For waterfowl expertise, I turned to my long-time friend and Smithsonian colleague Scott Derrickson. Zdenek Hubálek of the Czech Republic was my choice for virologist, based on the fact that he and his colleague, Jiri Halouzka, had published the most recent review paper on the relationship between migratory birds and West Nile virus outbreaks in Europe. Both Scott and Zdenek generously agreed to lend me a hand.

The resulting manuscript pretty much followed what I had said at the meeting, namely that if migrants were indeed the main reservoir and transporters among regions, either the disease would be carried quickly throughout the hemisphere during the fall 1999 migration season, or, since it appeared to be quite lethal to New World birds, it would simply die out as the mosquito population went into dormancy in October. Fortunately, however, I did not leave it at that. I decided that, like the real experts, I had better see if I could get input on our thoughts from the man himself, Dr. Reeves. So I tracked him down, gave him a call, and asked if he would consent to provide a critical review, to which he gracefully agreed.

His comments were extremely helpful, especially with regard to one aspect in particular. The draft of our paper leaned heavily on the West Nile studies conducted in western Europe, especially the summary paper by Hubálek and Halouzka in which they noted that the evidence that migrants were the introductory host causing European outbreaks was extensive and that the alternative, namely that active virus could overwinter in an area by being passed from mother to daughter in mosquito populations, was extremely unlikely. I included a statement to that effect in my paper for the CDC, but Dr. Reeves, who in fact had done some of the original work on overwintering of arboviruses in mosquito populations back in the '70s, advised me to tone my statement down. His research with a New World relative of West Nile showed that such persistence was entirely possible. He had found that infected female mosquitoes could pass active virus to their offspring, and that a small percentage of daughters were still viremic when they hatched in the spring. Thus, thanks to him, I left some wiggle room concerning the possibility that mosquitoes themselves could serve as the reservoir host for the disease, which turned out to be very important.

The West Nile chapter of my research career was behind me once the EID paper was out so far as I was concerned. Nevertheless, I kept an eye on what was happening with regard to the spread of the virus. After all, my coauthors and I had made some specific predictions based on knowledge of the volume, speed, direction, and destination of fall migrants potentially passing through the New York region. The CDC had alerted every public health department in the country to be on the alert for the virus by testing dead birds brought in to them, an immense effort but ultimately very effective in tracking virus movement. These data showed that, contrary to predictions, the virus had largely disappeared by October 1999 only to reappear on May 12, 2000, not far from the 1999 Queens epicenter. Reports for the 2000 season were equally unexpected, showing a gradual outward spread—east, north, west, and south—but not farther than a couple of hundred miles from the original outbreak site. The 2001 season showed more of the same; random expansion from Queens at a rate of about fifty miles a month during mosquito activity season (April to October in the northeastern United States).

I found this pattern to be very curious, given the assumed potential for movement of thousands of miles in a matter of days if migratory birds were indeed the introductory host. Still, I was not a virologist and would have done nothing more than ponder had I not been contacted by the Society for Applied Microbiology, which invited me to attend, at their expense, and present a talk at their annual meeting to be held in Nottingham, England. The paper would be published in their journal. I decided that this invitation gave me the excuse I needed to delve deeper into the evident movement of West Nile virus in the Western Hemisphere subsequent to its August 1999 appearance.

The meeting in Nottingham was different from the Fort Collins meeting in that West Nile virus was not the focus. Nevertheless, some of the top virologists in the world were in attendance, including some very familiar with West Nile virus in its Old World manifestations. As such, I thought it an excellent venue for airing my skepticism regarding the role played by migrants in transport of the disease, at least in its New World form. In my presentation, I showed the audience maps from the CDC website detailing the known spread of the disease from its 1999 New York epicenter, and I concluded that mosquitoes, not birds, had served as the reservoir host, allowing the virus to overwinter in the exact region where the original outbreak had occurred (as Dr. Reeves had suggested might be the case). I also opined that migrants had to date had little or nothing to do with the spread of the disease due to the very slow rate and randomness of movement, rather than rapid, north-south leaps, as one would expect if avian migrants were the culprits.

Zdenek Hubálek joined me as coauthor again on the resulting publication in the *Journal of Applied Microbiology*, which came out in 2003. That year showed more of the same type of movement of the disease (as documented by dead birds turned in to state public health departments and collated in tables and maps on the CDC website), but a rapidly expanding New World literature on the topic continued to focus attention entirely on migratory birds as the introductory hosts, and none mentioned the possibility put forward in our paper that avian migration was not involved, at least as yet.

The Ornithological Council was formed by the American Ornithologists' Union to serve as an advisory body to governments and the public on matters of policy relating to birds. During the Council's summer 2003 meeting, West Nile was raised as an issue. A member whose specialty was bird diseases presented a minilecture on the involvement of migratory birds in its spread, and advocated that the Council put out an information sheet to our governmental constituencies warning of avian involvement. I objected, explaining what the data actually showed, arguing that we needed to be extremely careful in what we claimed and recommended. In my experience, misguided government policy could do significant damage while serving no useful purpose. Other members of the Council agreed with me, and our public notice was modified accordingly. Nevertheless, it was this meeting that convinced me of the need to try to do something more to clarify what scientists actually knew regarding the role of migratory birds in movement of West Nile virus in the Western Hemisphere.

Subjecting the remarkable data set of bird death dates and localities collected across the country since 1999 by the CDC to mapping using geographic information systems (GIS), modeling, and statistical analysis seemed to me to be the best way to go. None of these talents was in my skill set, so I turned once again to putting together a team that had them: Dave King and Swen Renner for migrants and statistics, Peter Leimgruber and Jamie Robertson for GIS, and, most importantly, Brad Compton for the modeling. The resulting paper, published in the journal *Vector-borne and Zoonotic Diseases* (after being rejected by the *Proceedings of the National Academy of Sciences*), demonstrated clearly that migratory birds had little or nothing to do with movement of West Nile virus across North America during its first four years in the New World (1999–2003). Rather, it appeared that random dispersal by some viremic nonmigratory species, like the House Sparrow or even mosquitoes, provided the most parsimonious explanation for documented spread.

AVIAN INFLUENZA, H5N1

A similar tale explains my involvement with another pathogen, the H5N1 form of avian influenza. This virus was first identified from a flock of domestic geese in Guangdong Province, China, in 1996. First human deaths from the disease occurred in poultry workers in Hong Kong in 1997, but it was not until 2003 that the world health community really began to take notice as the pathogen spread quite rapidly across east Asia, apparently through legal and illegal transport of domestic fowl and their products, causing widespread deaths in chickens and geese. By 2005, the virus had been detected in domestic bird flocks across much of Eurasia, and although human incidents were mostly restricted to a relatively small number of people (less than two hundred), nearly all of whom worked in the poultry processing industry in Southeast Asia, a death rate of nearly fifty percent, caught the public's attention.

My involvement with West Nile virus had given me some idea of both the etiological and psychological complexities of infectious diseases. "Going viral" is a very real phenomenon, both figuratively and actually. My interest was stimulated in particular by the claim in publications on H5N1 of the probable involvement of migratory birds in spread of the pathogen. I examined the literature carefully and found very little evidence of migrant involvement. This information stood in stark contrast to testimony from some leading experts in the field of bird flu on US morning talk shows who predicted with a high degree of certainty that H5N1 would shortly make its appearance in the United States, thanks to its ready transmission by migratory birds. In addition, they warned, once the virus appeared here, it would likely undergo "reassortment," a process unique to viruses in which two different viral forms combine genetically in a way that produces a new virus with characteristics of each progenitor. The structure that this reassortment would take, they maintained, likely would produce a virus that could be readily carried by migratory birds but could also be spread to humans in highly lethal form, perhaps re-creating the terrible perfect viral storm generated by Spanish flu at the end of World War I.

Ingesting this prognostication with their morning coffee proved unhealthy for many people in the United States who became gripped by fear of what this disease might do to them personally. Although no human cases had been reported in Europe, let alone the New World, demands on our government were intense to stockpile huge numbers of doses of Tamiflu, a wide-spectrum antiviral drug, in anticipation of the pathogen's imminent arrival.

Symptomatic of this hysteria was a conversation I heard on NPR in which a talk-show caller stated that if migratory birds were the problem, why not just eradicate them?

Public prognosis by a few of our leading virologists of viral Armageddon fueled by migrants seemed unnecessarily inflammatory, as well as extremely unlikely given what was known regarding H5N1 etiology, so I contacted my old pal Zdenek Hubálek to get his views on the topic. After all, not only was he an expert on the relationship between migratory birds and virus movement, he also lived in a place where the virus had actually been detected, albeit in domestic fowl. Zdenek confessed that no viremic migrants had yet been found in Europe, although he expected them soon. Tens of thousands of waterfowl were being checked across the continent, and the trajectory for a new virus that is highly lethal to its host is likely to result in rapid mutation or reassortment to a less lethal, more readily transmissible iteration. Thus, he predicted that H5N1 would soon be found in migrants, but probably in a far less dangerous form. I asked him if he would be interested in writing another paper with me for publication in the CDC journal where we would lay out the facts and likely future of H5N1 in the planet in general and North America in particular. He agreed, and we set out to pull the paper together.

In the resulting article, we laid out the facts regarding avian influenza and its H5N1 version as known at the time of publication, which was at the height of the world epidemic in late 2006. We explained that avian influenza Type A viruses, of which H5N1 is one, are common and widespread in birds. Most such viruses attack the intestinal tract of the host and are spread mainly by shedding in host feces. Waterfowl are the most susceptible because they are often exposed to water contaminated by infected fecal matter, especially during postbreeding molt migration, when large flocks often occupy the same feeding zones for prolonged periods. A secondary mode of infection occurs when predators, such as hawks and owls, eat infected prey. Infection by most types of avian influenza is asymptomatic for the avian host (they don't get sick), and humans are not normally susceptible to infection by avian influenza viruses. Nevertheless, several subtypes of avian influenza have infected humans, and three of these caused major human pandemics during the past hundred years. H5N1 is an entirely avian influenza subtype at present. Humans can become infected, often with lethal results, but only by inhaling or ingesting massive doses from excreta or tissues of infected birds. Passage of the virus from human to human is very rare; there are only two or three known instances out of several hundred cases. The more humans become infected with H5N1, the greater the possibility that reassortment with a human influenza virus will produce a lethal form that is spread readily between humans. However, viral transmission strategies differ fundamentally for those that infect humans versus those that infect birds. Bird viruses show an affinity for the host's intestinal tract, and interhost transfer occurs mainly by fecal contamination of shared water bodies. Human viruses more often attack the respiratory system and depend on shedding in respiratory vapor for interhost transfer. If or when a reassortment or mutation of H5N1 were to produce a virus capable of transfer from human to human, the virus would not likely be particularly effective in transfer among birds. In other words, migrants likely would play little or no role in spread of such a virus.

We further explained that, while the main hosts and reservoirs for most forms of avian influenza viruses were migratory waterfowl and poultry (ducks, geese, and chickens), the original form of H5N1 caused high rates of illness and death among birds exposed. Sick migrants make poor introductory hosts, which presumably explains why no migrants were reported as infected from this form of the disease from the time of its discovery in 1996 until December 2002, when several migrant and exotic waterfowl were found dead at a Hong Kong park. Subsequently, of 3,905 outbreaks of H5N1 reported between December 2002 and February 2005, all involved captive birds or poultry. Up until August 2005, only two outbreaks had been confirmed in which migrants alone were involved (no apparent domestic fowl present)—one in China and one in Mongolia.

Thereafter, the situation changed somewhat and several infected migrants were discovered, perhaps signaling genetic modification of the virus to a less lethal form. Regardless of the cause, by 2006 some migrant individuals of some species had been found infected with live virus at various localities across Eurasia, although no human infection from a migrant host had been reported.

Given this situation (as of the middle of 2006 when the paper was written), we considered what might be possible modes of entry for the virus into the Western Hemisphere. As in the case of West Nile virus, we suggested three possibilities: normal interhemispheric migration, cyclonic storms, and legal and illegal imports. Entry into the hemisphere via migration or cyclonic storms, each of which would require flights of several thousand miles, seemed highly unlikely as modes of entry given their rarity and the likelihood that no sick bird could complete the flight, unless of course the virus modified to become completely asymptomatic, in which case it would likely pose no threat either to avian or human health. Thus, legal or illegal imports of live birds or infected products seemed the most likely mode of entry, especially as this mode is how the thousands of outbreaks elsewhere around the globe had occurred. None up to the time of publication could be traced to import by free-flying migrants.

We then considered what the appropriate public health response should be if an outbreak of H5N1 in the Western Hemisphere were to occur. We pointed out that it was important to remember that only birds could serve as introductory or reservoir hosts for H5N1 or any other form of avian influenza. Neither mosquitoes nor any organism other than birds could serve as introductory hosts so far as was known as of mid-2006. In summary, we maintained that movement of H5N1 as a viable infectious disease to the Western Hemisphere seemed to us improbable. If it were to occur, the likely mode of entry would be via legal or illegal import of infected fowl or fowl parts. If the disease were to enter in this way, it would likely have lethal effects on any birds subsequently infected, which would signal rapid quarantine and flock destruction by public health authorities, likely limiting outward spread from the site of entry, and even if the virus were to escape such quarantine and infect birds at other sites, the virus in its present form posed no danger to humans who were not exposed directly to infected fowl, their feces, or their parts. We further explained that while some had argued that reassortment of H5N1 could occur in which the disease became readily spread among humans yet retained its lethal nature, this could happen only if the disease were to infect a human already infected with a virus of a type readily transmissible among humans, presumably via respiratory effluvia. If that were to occur, migrants as introductory hosts would be out of the picture. Human to human would be the mode of transfer, and the disease could be readily shut down via quarantine of infected individuals.

Thus, we concluded, H5N1 was unlikely to enter the Western Hemisphere, and if it did, the chances were vanishingly small that it would pose a significant public health risk to the people of the United States, far less, indeed, than West Nile virus.

An online version of our article came out in the CDC journal *Emerging Infectious Diseases* in September 2006, followed shortly thereafter by the print version, at a time when the Eurasian epidemic was about at its peak, and avian influenza alarm was at its most intense in the United States. The only public reaction that I was aware of came from "Elizabeth," a primary care physician from New Jersey. She emailed me to question my certainty regarding the low level of threat from the disease to US public health. I responded in broadly reassuring tones, reiterating the main points of our paper. Her subsequent email stunned me. She said that I was naïve, stupid, evil, or all three, concluding with the statement quoted above expressing her hope that I would be the first to suffer from the pathogen's effects when it did arrive in this country. Which seems excessive. Even most people who know me wouldn't wish for that.

A few virologists and ornithologists were also unhappy with our prognosis, as expressed by negative comments in their publications. It is hard to know exactly why, but it is undeniable that millions of dollars in grant funds from the National Science Foundation (NSF) and the National Institutes of Health (NIH) had become instantly available for migratory bird and H5N1 studies when the lethal disease's New World arrival appeared imminent—money that would vanish as fast as spit on a hot griddle if the threat were to diminish or disappear.

It seems to me that the ways in which these two epidemics were dealt with contain some salutary lessons going forward. In the case of West Nile, leadership allowed science to guide policy. A huge collaborative effort organized by the CDC immediately after the pathogen's appearance in the New World brought together experts in all of the known factors affecting the etiology of the disease. This gathering resulted in succinct summaries of what we did and did not know, and it established country-wide actions in terms of mosquito control, monitoring the spread through dead-bird surveys, advising people of what precautions to take for preventing mosquito exposure, and rapid screening of patients showing suspicious symptoms. Political leaders were reassured and, for the most part, gauged their agency actions and advice to the public to the actual level of threat.

Even though West Nile hysteria in the New York region had reached a fairly high pitch by October 1999, with thousands of people infected—sixty-two of whom required hospitalization, and seven of whom had died—the extraordinary skill demonstrated by the CDC in orchestrating a public health response precisely geared to the scope of the threat kept people fairly reassured. There were, nevertheless, some conspiracy theories voiced. The *New Yorker* published an article speculating that Saddam Hussein or some similar terrorist leader might have tried to use West Nile as a bioweapon. This suggestion was taken sufficiently seriously that I was contacted by a

US Senate staffer and a Defense Department analyst for an assessment of this possibility. I explained in written responses to both that the likelihood seemed remote, despite the fact of the virus's evident Middle Eastern origin. The seemingly perfect storm required for a devastating outbreak—viremic introductory avian host, huge numbers of avian amplifying hosts in the immediate vicinity of vast swarms of potential vector mosquitoes located near large human population centers, just seemed too far out for a decent weapon system.

The contrast with how the H5N1 avian influenza outbreak was handled could hardly be more stark. Doomsday scenarios were rapidly promulgated and accepted as fact in the United States based on highly speculative and unlikely hypothetical possibilities that were contrary to all that was known regarding the actual etiology of the disease. This hysterical reaction was quite remarkable, especially when compared with reactions in countries like Vietnam where scores of people died. In these places, actions were gauged specifically to the known threat: domestic bird flocks were screened, and, where infected birds were found, their flocks were quarantined and destroyed. Poultry workers were provided with gloves for handling the birds, and consumers were advised to do the same, as well as to be sure to cook the products thoroughly. These actions brought the epidemic quickly under control in most countries where it was found.

According to the World Health Organization (WHO), the total number of deaths up to 2018 attributable to H5N1 since its 1996 appearance was 454, over half of which occurred in two countries, Egypt and Indonesia, where public health response was weakest. No human infections or deaths were ever recorded for Europe or the New World (except for one death in Canada resulting from travel to China) and only one in Africa outside Egypt. No one died of H5N1 anywhere in the world during 2018. In addition, no migratory or domestic bird had been found carrying H5N1 in the Western Hemisphere despite the screening of tens of thousands of birds. Viremic migrants have been found in the Old World, but lethality of the disease for birds has declined significantly.

DISEASE THREATS TO MIGRANTS

Migratory birds were key players in both of these stories, as they will be in the future. Their ubiquity and potential for pathogen transport ensures their place among likely suspects in the etiology of any new pathogen. Nevertheless, actions should be based on facts, or, where those are lacking, on the most likely probabilities. In addition, of course, there is the question of what effect new pathogens might have on migrant populations themselves.

When West Nile virus first appeared in the Western Hemisphere in August 1999, it appeared to be deadly to any bird that contracted it. As I have mentioned, for the first years of its presence here, its movement across the continent was tracked by picking up birds that had died from the pathogen. Wildlife specialists worried that mainland bird populations might be threatened by the outbreak, especially among species that tended to form large flocks for roosting in fall and winter, like crows. Several papers were published addressing this question and purporting to find evidence of local decimation in some populations of some species. However, the National Breeding Bird Survey showed no evidence of decline among those species considered most vulnerable, namely corvids (crows and jays). Presumably this was because both the virus and birds changed; the virus likely changed either through mutation or reassortment to a less-deadly form, and bird populations developed resistance. Nevertheless, the history of new diseases is indicative of their potential to pose conservation problems for migratory birds as well as other aspects of the biota. Species with large, widely distributed populations likely will survive and recover from epidemics. However, those with small populations certainly are of concern. The probability that they will survive is much lower.

Hunting

> Sing a song of sixpence,
> A pocket full of rye,

> Four and twenty blackbirds
> Baked in a pie.
> —"Sing a Song of Sixpence"

Over the history of European colonization in North America, unregulated hunting has posed the greatest threat to migratory birds. The millinery trade (herons and egrets), market and sport hunting (waterfowl, shorebirds, and upland game birds), and misguided predator control (eagles, falcons, hawks, and owls) all played a roll. By the early 1900s, excellent regulations had been put in place so that now these chronicles serve as mere cautionary tales regarding the potential devastation that harvest of migrants can cause. At present, there appears little cause for worry concerning hunting as a conservation factor in the United States despite recent relaxation of "taking" rules for the Migratory Bird Treaty Act by the Trump administration. Most other countries have no such laws or no enforcement in those that do have them. Where subsistence hunting still exists, the potential for eradication of migrant populations is limited only by the means available for harvest.

Cats

> Tyger Tyger, burning bright,
> In the forests of the night;
> What immortal hand or eye,
> Could frame thy fearful symmetry?
> —William Blake, "The Tyger"

Felines are a fraught subject. As predators of humans, they haunt our dreams, and even small ones, like the domestic house cat (*Felis catus*), with their stealthy ways and vicious attacks on, and torture of, their prey, inspire superstitious dread in some of the more credulous members of society. And yet, the relationship between many humans and their cats has been a source of comfort and even veneration for millennia. Thus, it is perhaps not surprising that cats as a conservation issue are at the center of an extraordinarily heated controversy, exemplary of the value chasm between conservation and science.

The house cat derives from the Eurasian Wild Cat (*Felis sylvestris*) (now extinct in the wild), whose domestication occurred over ten thousand years ago in the Middle East (Tutankhamun likely kept one as a pet). The species was introduced into the New World at the time of colonization. We have our own wild cats here, of course—the cougar, lynx, and bobcat north of the tropics and several more from Middle and South America. Nevertheless, domestic cats found life in the vicinity of human habitation quite congenial throughout the hemisphere even when on their own, and they have become a very successful example of how an introduced species can prosper in new environments.

The question at issue is, *Do free-ranging cats threaten continental populations of any Western Hemisphere species of bird or other vertebrates, and, if so, what should be done about it?* Pete Marra, my former colleague at the Smithsonian (now director of the Georgetown Environment Initiative, Laudato si' professor in biology and the environment, and professor in the McCourt School of Public Policy), and his coauthor, Scott Loss, had this to say in a recent (2017) article on the topic: "Free-ranging domestic cats are increasingly the focus of policy and management attention as well as controversy due to their substantial environmental impacts and their popularity as pets." Their position (and that of many others) is that cats kill billions of birds in the United States and elsewhere each year and cause serious damage to populations of many of these species. According to William Lynn and colleagues in a recent paper published in the journal *Conservation Biology*, Loss and Marra "equate citizens, nongovernmental organizations, ethicists, and scientists concerned about the well-being of free-ranging cats . . . with science deniers, a term used to describe those who mislead the public over the harms of smoking, ozone depletion, and climate change."

This exchange between putative scientists on opposite sides of a question is a perfect example of why I do not consider "conservation" to be a science. In normal scientific discourse, at least since the time of Francis

Bacon (early 1600s), the data are supposed to speak for themselves. The scientist presents a hypothesis and puts forward the information gathered in testing it in order to justify a conclusion in which the hypothesis is either accepted or rejected (requiring formulation of a new hypothesis). Clearly that is not the approach taken by the various individuals and organizations on opposing sides of the cat controversy.

What we see instead in these debates is the marshaling of every form of argument that might be hoped to sway a neutral person to the expounder's point of view. This form of presentation is not science. However, it is perfectly valid in policy discussion. Samuel Johnson, eighteenth-century author of the first comprehensive English dictionary, was one of the world's great sophists. He maintained that, while presentation of the merits of the argument should take precedent in a debate, it was perfectly OK to use any other pejorative verbal weapon that came to hand, including, but in no way limited to, the stature (physical or societal), gender, or ancestry of one's opponent. As an example, had he been my opponent in the cat debate, he might have argued, "The pathetic, hysterical, and confused arguments presented by my pied, diminutive, nearly hairless, Irish friend demonstrate that he has spent too long closeted with inscrutable and arcane codices to be able to present a coherent case." A bit harsh, you might say, but with just enough elements of truth to give a neutral observer pause (except maybe the "hairless" part—my comb-over covers pretty well, and besides, I mostly wear a baseball cap). All's fair in love, war, and policy disputes.

Viewed in this light, one can see that the attack wielded by Loss and Marra against the various individuals and groups questioning the negative effects of cats on birds fall well within the grand disputatious tradition. As Dr. Johnson maintained, the only point of a conversation is to win in the eyes of those observing it.

Nevertheless, the cat debate includes some pretty heavy stuff, from both sides, that deviates a long way from the scientific question at issue, namely do cats reduce continental populations of any bird species, or more specifically for us, any migratory bird species?

The fact that cats kill birds is undeniable, but they prefer small rodents,

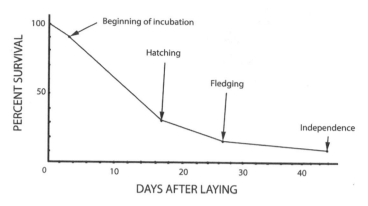

FIG. 11.1. Percent survivorship for Prairie Warblers from day of laying until day of independence from adults. BASED ON DATA FROM NOLAN 1978

especially mice. And many things kill birds: hawks and owls kill birds; mice, squirrels, and chipmunks kill birds (mostly in the egg stage); opossums, skunks, raccoons, and snakes kill birds (again, mostly in the egg stage); jays and crows kill birds; cars kill birds; windows, TV towers, and ceilometers kill birds; storms, wind, and weather kill birds; and so forth. All birds must die, and most die within days of leaving their mothers' oviduct, as shown in Figure 11.1, which is based on Val Nolan's work on Prairie Warblers. By the time young Prairie Warblers are ready to leave the nest, about three weeks after laying, over eighty percent of them are already dead, a statistic comparable to that of most other open-nesting songbird species. This observation helps to place mortality factors for adult birds into some reasonable population context. Nestling deaths exceed all known causes of adult deaths by a factor of four. Put another way, chipmunks, a major predator on small, open-nesting songbird nestlings, likely are responsible for more deaths than buildings, cats, pesticides, and wind turbines combined in these species.

Given the myriad factors accounting for bird deaths, one must ask why it is that domestic cats are singled out for special vilification and national efforts at control. I suspect that it could be because cats are more often observed in the actual act of killing, and they seem to relish the helplessness of their victim. In addition, cats have a well-earned reputation, having been responsible for extinction of island populations of flightless birds and other small terrestrial vertebrates. But are they a threat to any mainland population of migratory birds?

In considering this question of fact, it is important to think about the ecology of the cat. Like many introduced species, their preferred habitat is usually in the immediate vicinity of human habitation—towns, cities, and farms. They are seldom seen far from such areas, partly because rodent populations are higher near people, but also because, in most native habitats, cats don't compete well with indigenous predators (like raptors, bobcats, and foxes) for small-mammal prey, and also because cats are vulnerable to predators like Great Horned Owls and coyotes.

Even in the disturbed habitats that they prefer, cats' principal prey is not birds. Cats are terrestrial predators. Although prey type varies by season, seventy percent or more of their prey are small mammals. Their main avian prey are ground-frequenting birds like doves and quail or recently fledged young of a few species that are abundant in residential areas, such as robins and House Sparrows.

The data on cat damage to avian populations has two principal sources: (1) extrapolation from *observed* cat kills to the universe of *assumed* cat kills, and (2) carrying capacity theory regarding additive versus compensatory mortality. Extrapolations take the relatively small number of observational studies of actual cat kills (a few dozen?) and use the estimated total number of free-ranging cats in the country as a multiplier. As one would expect, resulting values differ wildly, but Scott Loss, Tom Will, and Pete Marra, in a 2013 paper in the journal *Nature*, concluded that about 2.5 billion birds were killed annually by cats in the United States. They make no mention of the kinds of birds involved.

The second major approach to assessment of cat damage to avian populations is use of the concept of compensatory versus additive mortality. These terms were proposed in the 1970s for the purpose of helping wildlife biologists to set harvest or "take" numbers for game species that would not harm their populations. In essence, they were a way of applying the idea of carrying capacity to practical issues of game population management. The theory states that for any population of game, say deer, if that population is at carrying capacity, then the number of offspring produced in a given year that will join the breeding population in the next year will only be equal to the number of adults that die in the interim. The rest of the offspring will die anyway because the habitat is fully occupied and will not support them. What they die of is insignificant—hunger, disease, predators, hunting—it doesn't matter. These kinds of mortality are called "compensatory" in that if a deer dies from being shot instead of hunger, it will have no negative effect on the population. The shooting death just "compensates" for mortality from some other source. However, if *more* deer die from hunting (or any other single cause) than the habitat can support, the effect on the population is "additive," meaning it will "add" to total deaths beyond those expected to die as a result from carrying capacity limits, thus lowering the population numbers below carrying capacity. Therefore, population modelers advise that in setting the harvest level for any game species, care should be taken that the number of individuals taken does not bring the population below carrying capacity. If you repeatedly allow hunters to take more animals than the habitat can support, you will wipe out the species, at least locally or regionally, which is what happened in the eastern United States for deer and many other game species up until the early 1900s when game laws first became federally mandated.

So how does this apply to cats and birds? The argument put forward by Loss and Marra is that we don't know whether bird mortality from cats is compensatory (that is, that the birds killed by cats would have died anyway from one cause or another because more offspring are produced than the habitat can support) or additive (in the sense that continued cat depredation will cause avian population decline or even local or regional extirpation). The problem with this reasoning is that it is false. Since there are no data demonstrating continental population decline as a result of cat depredation for any bird species, migratory or otherwise, the "compensatory versus additive" discussion is irrelevant. Loss and Marra more or less acknowledge this fact, but they claim the problem is the complexity of the issue. It is very difficult to determine whether declines are occurring, and, if they are, what cause or causes are most important. Therefore, they conclude, since we know cats kill a lot of birds, and we know how to control that issue (by taking cats out of the wild), it only makes conservation sense to do so to the degree possible.

So, the answer to our question of cat guilt in continental bird decline is straightforward from a scientific perspective. At present, there are no data demonstrating that cats constitute an existential threat to any species of North American bird, migratory or otherwise.

The reader should not take my presentation to mean that I favor one side of this policy argument over the other. In fact, I am agnostic. Cat control is a conservation issue, and, as such, is up to local jurisdictions to decide what to do based on the opinions of their constituents regarding what is most important to the majority—fewer bird deaths due to cats or more cat freedom. You decide. Either way, available data indicate that migratory bird populations will be fine, at least in terms of cat effects.

Pesticides

> The birds . . . where had they gone?
> —Rachel Carson, *Silent Spring*

As mentioned above, DDT and similar poisons were responsible for widespread decimation of populations of several avian predators, including most spectacularly Brown Pelicans and Peregrine Falcons, whose US populations were driven nearly to extirpation before environmental regulations brought use of this and similar poisons under control. Vigilance is necessary to make certain that these kinds of toxins are not allowed because there is always industry pressure for their use.

As an example, I was asked to participate as an expert witness in July 2004 before a panel convened by the National Academy of Sciences to consider lifting the ban on manufacture and distribution of DDT to third world countries for the purpose of mosquito control.

Why I was selected to defend the ban is open to conjecture. I have never published on the topic, but I had worked with Academy staff on some other issues, plus, as a Smithsonian employee, I was available. Driving into Washington, DC, from Front Royal, Virginia, to attend the panel would be cheap for them and easy for me. Why my opponent before the panel was chosen is less difficult to understand. He had been a star witness for the agricultural industry in the congressional hearings considering the DDT ban in 1972; he was articulate, colorful, and a "scientist" (chemistry professor, I think).

I was extremely nervous on the day of my testimony before the panel, which was composed mostly of infectious disease specialists and physicians—the main argument for lifting the ban was to allow DDT use in Africa to help control malaria. I had twenty minutes to present the case against lifting the ban and was to be followed by my battle-hardened adversary. I have few illusions regarding my effectiveness as a public speaker, and I worried that I would "let the side down."

My concerns, however, were overblown. The star didn't show. Although long-scheduled to "fly in from the Coast," he sent word that the date conflicted with a planned hang-gliding vacation with his girlfriend or some such nonsense. Personally, I think it dawned on him that a National Academy review panel is not a congressional hearing.

Basically, a congressional hearing is Kabuki theater. Each member of the cast plays a character well known to the audience: the Bank Lobby Flack, the Bleeding-Heart Environmentalist, the Corporate Friend, and so forth. During the 1972 hearings, my erstwhile opponent played the Industry-Supporting Scientist. He probably realized that appearing before a group composed of experts in their field was a different kind of game altogether.

The argument in favor of lifting the ban (aside from economic boon to US chemical companies) was that it was a cheap and effective method for local control of the mosquito populations responsible for spread of malaria and other deadly diseases. The arguments against were that there were other equally effective methods for mosquito control that did not involve poisoning the environment. In addition, thirty years of research on the chemical had not improved its known standing as a brain poison that short-circuited acetylcholinesterase-modulated neural transmission for insects, vertebrates, and many other animal groups.

Fortunately, the Academy panel was convinced that the risks outweighed the benefits, but the very fact that the panel had been convened should give us pause. The Academy associate responsible for preparing background documents had readied a summary supportive of the industry position and had lined up the witness mentioned above, who had testified to Congress about the benefits of DDT and lack of evidence that it caused environmental problems.

Wind Farms

> For look there, friend Sancho Panza, where thirty or more monstrous giants present themselves, all of whom I mean to engage in battle and slay, and with whose spoils we shall begin to make our fortunes; for this is righteous warfare, and it is God's good service to sweep so evil a breed from off the face of the earth.
>
> —Miguel de Cervantes, *Don Quixote*

Windmills kill birds. That fact is undeniable. However, with the possible exception of Golden Eagles in California, there is no evidence to my knowledge that these structures have had an effect on bird populations. The most worrisome wind farms are the huge ones being placed along major migration routes, like the western coast of the Gulf of Mexico in Texas and on the shores of the Great Lakes. These installations are located along the most densely traveled migratory routes in North America, traversed by members of hundreds of species of birds, several of which have small populations considered threatened or endangered in the states where they breed. Most birds fly above the height of wind mills during migration, but weather conditions (low cloud ceilings) can put them at risk. Bats are much more susceptible. They migrate at lower altitudes, and populations of the eight species that travel these corridors are much smaller. More work needs to be done to assure that no bird populations, migratory or otherwise, are threatened by wind harvesting.

Breeding Season Threats

> The sedge is withered from the lake,
> And no birds sing.
> —John Keats, "La Belle Dame sans Merci: A Ballad"

Destruction, fragmentation, pollution, or any of the other factors that can make breeding season deadly for a migratory bird are high on the list of conservation concerns in the minds of most land managers. After all, such changes often are the most visible, and the fact that if the breeding habitat is gone the birds are gone is undeniable. Still, as discussed in Chapter 8, the migrant life cycle makes determination of exactly when and where (or even, if) populations are threatened difficult. In some cases, such as the loss of prairie potholes to industrial agriculture in the northern Great Plains of the United States and Canada, the negative effect on several species of migratory waterfowl that used them for reproduction seems indisputable. Other situations are less clear-cut, as discussed below.

Breeding-Ground versus
Wintering-Ground Habitat Limitation

I have mentioned the precipitous drop in long-distance forest-related migrant populations as measured by the National Breeding Bird Survey in the 1980s and 1990s, which, at the time, was attributed by most to breeding-ground habitat fragmentation assumed to be occurring across vast areas of the country. We were skeptical, as discussed in Chapter 8, arguing (in the literature) that winter habitat loss could be the culprit, but there was really no way to resolve the dispute empirically.

It took us fifteen years, but eventually colleagues and I were able to do a project to actually measure total amounts of habitat and potential population size in both breeding and wintering areas for a migratory species. It wasn't easy. The saga begins back in 1985 when I was working as a research scientist at the Caesar Kleberg Wildlife Research Institute at Texas A&I University (now Texas A&M University, Kingsville). Carol Beardmore, who was charged with endangered species management for Region 2 of the US

Fish and Wildlife Service (USFWS), contacted me to ask if I would serve on the Endangered Species Recovery Committee for the Golden-cheeked Warbler. I readily agreed, although I had no first-hand experience with the bird and had never seen one in the wild.

The goldencheek breeds only in central Texas and winters in Central America, and at the time, it was considered to be in serious danger of extinction mainly due to loss of its oak-juniper breeding habitat to land clearing for homes, pasture, and agriculture. The purpose of the committee was to review all available information on the species and prepare a report suggesting a target minimum population size for the bird to assure its preservation. We were also asked to create a prioritized list of conservation actions to help achieve that goal. At the first meeting, I learned that, with the exception of a few graduate student theses, nearly all work on the bird was being done at Fort Hood under the auspices of a well-staffed, well-funded (over a million dollars a year) nest-survey and banding program. My first question was, "What do we know about the wintering ground?" The answer was, very little. In fact, at the time there were exactly twenty-nine published accounts of wintering-ground occurrence. I suggested that even if the only problems facing the bird were in Texas, we should at least know something about its life during the rest of the year.

Eventually, after several years of pestering on my part, USFWS gave in and said I could submit a proposal for a grant to do the work. By that time, I had moved to the Smithsonian Institution's Conservation and Research Center (now known as the Smithsonian Conservation Biology Institute) and had set up a GIS lab with a couple of staff. With this new capability at hand, I proposed to go down to Central America, figure out what habitat the bird lived in and at what density, and then to use satellite imagery and GIS programming to identify and calculate total amount of that habitat across its entire winter range. Once the project was funded (1995), I set about choosing a locality in which to begin. I chose Honduras ultimately, because of Burt Monroe's masterful treatment, *A Distributional Survey of the Birds of Honduras*. Mario Ramos's contacts through the Neotropical Ornithological Society put me in touch with the leading ornithologist in the country, University of Honduras biology professor Sherry (Pilar) Thorn, and in October I flew down to meet Sherry and to see if I could find the bird.

Wintering habitat for the goldencheek, according to Monroe, was highland (above four thousand feet) pine-oak woodlands, and the nearest locality to Tegucigalpa (twenty miles) where the bird had been found was a place with the euphonious name of "El Cantoral." On October 19, 1995, I rented a beat-up Datsun coupe and, with Sherry's son, José, riding shotgun, headed up into the rugged mountains north of the sprawling capital. Once we had turned west off the main highway, the "road" to the tiny settlement was no more than a cart track, but we encountered no ruts that José couldn't push the little vehicle out of, and by one in the afternoon, I was walking through disturbed (burned, grazed) but decent pine-oak at about five thousand feet.

Non-field biologists may find my terror at this moment difficult to comprehend. For ten years I had been blistering Fish and Wildlife and the goldencheek committee for doing nothing to learn about the eight months of the year when the bird was not in Texas. Finally, they had capitulated, and given me fifty thousand dollars to study the life history, population ecology, and distribution in its reputed winter range (mountains of Sierra Madre from southern Mexico to northern Nicaragua). Now, here I was, exposed and alone, figuratively poised over the bull's horns, two thousand miles from home in the classic "I can do that—how am I going to do that!?" moment. Monroe had said the bird was "rare" in Honduras. He had reported only two localities for the whole country, and no one had seen it at El Cantoral in sixty years.

Most birds in this habitat at this time of year in the Central American Sierra associate in mixed-species flocks composed of one to a few individuals of eight to twenty different species. The first flock I found consisted of the Grace's Warbler, Yellow-throated Warbler, Black-and-white Warbler, Painted Redstart, Eastern Bluebird, Western Wood-Pewee, Hepatic Tanager, and Chipping Sparrow, plus five Hermit Warblers and four Black-throated Green Warblers. These last two species are close relatives of the goldencheek, and each Black-throated Green in particular had to be carefully examined because females of the two can be quite similar.

A half mile farther on I found a second flock containing many of the same species plus the Unicolored Jay, Buff-breasted Flycatcher, and four Townsend's Warblers (another close relative of the goldencheek). The third flock, located five hundred yards or so from the second, contained several Hermit, Black-throated Green, and Townsend's warblers, Painted Redstart, Blue-headed Vireo, Spot-crowned Woodcreeper, Wilson's Warbler, *and three male Golden-cheeked Warblers!* Whew!

So that's how it started. With advice from my dear friend and Forest Service research scientist, Dick DeGraaf, I hired Dave King to head up the field work effort, assisted by several students of Sherry's when needed. Peter Leimgruber and Jeff Diez did the GIS and remote sensing. After three years, and hundreds of field hours in Guatemala, Chiapas (Mexico), and El Salvador, in addition to Honduras, we had mapped the Golden-cheeked Warbler's distribution and felt we had a good handle on its winter biology. At about the same time that our Central American work was being done, the USFWS funded a GIS/remote sensing mapping effort for the entire Texas range. For reasons best known to the organization, these data were never published, but we were able to obtain them from the contractor (David Diamond). Using this information in combination with our own, we were able to calculate total amount and distribution of breeding and wintering habitat, and to calculate carrying capacity (total number of birds that could be supported) for these two parts of the species range.

The results, published in 2003 in the journal *Ecological Applications*, were stunning: the Texas breeding range had sufficient habitat to support an estimated *230,000 birds*, whereas the Central American winter range had sufficient habitat to support only about *34,000 birds*. These data contain a clear conservation message: population size for the Golden-cheeked Warbler is controlled by wintering-ground habitat availability.

This effort at carrying capacity estimation for both breeding and wintering ground in a migratory species is unique to my knowledge. Nevertheless, its import has broad implications for a number of other migrants whose populations are assumed to be threatened by breeding habitat loss (such as Kirtland's Warbler and Bobolink). It is necessary to examine the entire life cycle carefully before drawing conclusions or spending a lot of money on conservation programs likely to have little or no effect on species preservation.

Breeding Habitat Loss: The Big Picture

When thinking about the effects of breeding habitat loss on migrants, it's important to consider the situation from a theoretical perspective based on the ideas on origin of migration presented in Chapter 9. From a deep history standpoint, we know that breeding habitat for many or most migrants has been completely obliterated every forty-one thousand to one hundred thousand years over the past 2.4 million years, only to be reoccupied at comparable intervals. This observation raises the question of whether breeding habitat loss is a conservation issue at the population level. In other words, even if breeding habitat were to disappear completely, many migrant species would likely continue to exist as nonmigratory residents on their wintering grounds.

As discussed above in the section on hunting, there is extensive evidence to indicate that this supposition is correct, at least for species known to have breeding populations in the tropics as well as the temperate zone. The clearest example comes from several species of herons and egrets whose temperate breeding populations were extirpated in the late 1800s by the millinery trade. After an absence of a century or more, the breeding populations of Snowy Egrets, Great Egrets, Little Blue Herons, and so forth have become reestablished throughout much of eastern North America as far north as New York and southern Ontario. The implication is that if a source population persists as breeders in the tropics, and the habitat exists in the temperate zone, a migratory population will develop through dispersal.

About half of Nearctic migrant species have tropical resident breeding populations. Results of the massive extirpation experiment performed on herons and egrets indicate that, if temperate breeding habitat exists, tropical resident populations will supply the dispersers necessary to reoccupy it as migrants. But what if there is no tropical resident breeding population, as is the case for half of our migrants? Would a tropical resident population reappear if breeding habitat were to disappear completely? From a theoretical point of view, I think it would. I would guess that a small and variable percentage of migrants remains to breed in their wintering areas each year after their cohorts have left for temperate breeding areas, and that number would increase dramatically when temperate habitat became scarce or absent. This idea, however, is entirely conjectural at this point. I have no data from any species to confirm its possibility.

Migration

> Boy, you in a heap of trouble.
>
> —Sheriff Joe Higgins to an out-of-stater
> stopped for a minor traffic violation, 1970 Dodge commercial

Surely, there are plenty of things to worry about as a transient—storms, predators, your next meal, unfriendly residents, and so on. Indeed, my Smithsonian colleague Scott Sillet has argued that the process of migration is the single highest cause of mortality for its participants. Maybe so. It's hard to argue with the fact that, in a stable population, if there were two robins breeding in your backyard last year, there will only be two robins again this year. Something had to happen to the ten babies or so they produced. Keep in mind, though, that sixty percent of Hooded Warblers came back to their winter territories in Tuxtla rain forest in the fall of 1974. That's probably pretty good odds for a half-ounce bird traveling four thousand miles, round-trip, and that includes survival through the breeding and postbreeding seasons as well. Migration may not be as dangerous as has been portrayed, at least once you've made the trip. There's a lot we don't yet know. And, it's important to remember that, if resources were not superabundant along the migration route during the transient period for all of the species involved, there would not be any migration.

In any event, migration is considered an important conservation issue at present, with focus placed on "flyways" and the likely stopover sites for transients along them. A "flyway" is a probability concept developed by waterfowl biologists for the purpose of planning the location of wetland refuges in places that would benefit the largest number of ducks and geese passing through on migration or spending the winter. Excellent work by field biologists, such as Brian Harrington, Raymond McNeil, Theunis Piersma, and Pete Myers, demonstrated that flyways existed for shorebirds, as well, and that stopover sites, like Delaware Bay, might be vital to some sandpiper populations, like the Red Knot.

Recently, efforts have been made by the Audubon Society to expand the concept to include conservation efforts for *all* avian transients, based on ideas drawn from the Migratory Route hypothesis derived from the *Zugunruhe* experiments described in Chapter 9. You will recall that, according to this conjecture, the migratory route is a genetic program, deviation from which can lead to disaster. If so, conservationists argue, then potential stopover areas along the route must be absolutely vital for survival, and they must be preserved. From one perspective, there is nothing bad about this idea whatsoever. Preservation is preservation. If it isn't necessarily vital for birds, other organisms will benefit. However, there are potential conservation downsides in that it can result in misidentification of where the actual threat to a population lies, just as preservation of Texas breeding habitat has no value for Golden-cheeked Warblers so long as nothing is done to preserve wintering habitat. With regard to stopover, there is extensive evidence, as I have shown, to indicate that migration route is *not* fixed. Which means that loss of stopover habitat in one place just means that the birds will go to another, taking a completely different route if necessary. Birds of many species do not have that option so far as the wintering ground is concerned.

Migration is all about food. Dispersal is the ultimate cause, but it is food that makes movement possible. Just like a 747, the migrant has to have fuel for the trip. I have discussed above several of the hazards confronting migrants during their journey, but it is likely that lack of food is the number one cause of death during migration. The reason that this situation is not recognized is that death from starvation most often occurs unseen by human observers, or it is attributed to some other factor when people do see dead migrants.

For instance, I have discussed the tendency for many male songbirds to leave the tropics for their temperate or boreal zone breeding territories a week or more before females. That such early migration has hazards was pointed out over a century ago by W. W. Cooke in his classic summary on migration timing and routes, *Bird Migration*, wherein he notes that late spring cold snaps in US temperate regions kill thousands of migrants occasionally. Most observers attribute this phenomenon to migrant inability to withstand cold temperatures, which is almost certainly not true. Birds, as I have mentioned elsewhere, are extremely efficient thermoregulators. However, in order to be able to survive under such conditions they must have food, and it is food, in the form of insects and other arthropods in their various life stages, that cold temperatures cause to disappear.

A second example of migrant vulnerability to sudden changes in food sources along the route was in the news quite recently (September 2020). Researchers from New Mexico State University reported thousands of dying and dead migrants of several western species, including swallows, flycatchers, and warblers, at various sites around the US Army's White Sands Missile Range. The vast fires in California and the northwestern states were deemed the ultimate cause, which seems likely, but what actually killed the birds was open to debate with some suggesting smoke inhalation and others chemical poisoning. I suggest that lack of food is the probable cause of death.

The sequence of steps taken by migrants in preparation for a migratory flight is shown in Figure 5.3. A fascinating finding from our work on transient Northern Waterthrushes in Texas, which has also been found for caged migrants, is that if no food is available, then the birds leave anyway (or try to in the case of caged birds). The reason is obvious. Why stay if you can't find food? But how is travel physiologically possible? The answer, discovered by work with birds in the laboratory, is that birds can use muscle as fuel for migratory flight if forced by circumstances. Thus, if the food necessary for laying down fat reserves in preparation for migration has been destroyed by fires, the migrants still have the option of leaving, the problem being that food had better be readily available when they land in their weakened state or they will die. Evidently it wasn't at White Sands (and many other places in the Southwest), so they died.

The most readily observed situation in which migrants run out of fuel in the middle of a flight occurs along the western coast of the Gulf of Mexico during spring migration. I have mentioned in Chapter 5 that many migrant species that migrate directly across the Gulf in fall take a much more westerly route in spring. At least one reason for this difference in northbound versus southbound routes is the probability of confronting fronts with strong north winds while over the Gulf in spring. When this situation occurs, the flight time over water can be extended by several hours, causing many of the voyagers to run out of fuel before they reach land. If you should happen to be standing on the Padre Island beach when this happens, as I have been on occasion, you will experience a most remarkable sight: thousands of migrants of many different species flying a few feet above the waves in a desperate attempt to reach the sand before their "tanks" are completely empty.

Wintering Season Threats

> What ecological changes are doing to Kirtland's Warbler
> on its wintering grounds in the Bahamas, where human
> populations are doubling in less than 20 years, should
> be a matter of grave concern to those concerned with
> perpetuation of the species.
>
> —William Vogt (Buechner and Buechner 1970)

I have mentioned that William Vogt convened a symposium at the Smithsonian to address his worries about the possibility that loss of winter habitat in the tropics could threaten temperate zone migratory birds. Few biologists at the time were concerned by the prospect. Our field work and theoretical considerations, as discussed in many places in this book, have shown that Vogt's concerns were well founded. The wintering season is the time of greatest vulnerability for most migrant species, in my informed opinion. However, field investigation into the details of all aspects of the annual cycle is really the only way to determine what actions need to be taken, and a species-by-species approach is always best.

Coffee

The program to promote consumer use of shade-grown coffee for migratory bird conservation provides a cautionary example. The late Russ Greenberg, my esteemed Smithsonian colleague, came out to our research center in Front Royal at my invitation to give a talk in the late 1990s on his research on migrants in Chiapas, Mexico. He focused on his work in traditional coffee plantations, making the point that such agricultural usages actually were entirely consistent with migratory bird conservation. Surveys performed by Russ and his students demonstrated that migrants wintered successfully in traditional coffee plantations (which involve growing coffee in shade, as undergrowth in forests) at levels comparable to those found in primary forest. He said that, in his capacity as director of the Smithsonian Migratory Bird Center, he planned to develop a program promoting the purchase of shade-grown coffee by US consumers as an important conservation activity—something the average person could do to help their warblers, vireos, and thrushes come back to us up here in the northland.

As I listened, I became concerned. I had two research projects relevant to Russ's idea. The first was with agronomist Charlie Russell back in the '80s when I was at the Caesar Kleberg Wildlife Research Institute. I had some money then to investigate ways to promote intercropping as a way to reduce the need for subsistence farmers to be continually clearing vast new swaths of forest. The idea was that, by growing the proper mixture of crops, nutrients would be replaced while providing both food for one's family and a cash crop that could be sold. Coffee was a part of our investigation.

We found that there are basically three ways that coffee is grown throughout the world: (1) as undergrowth in forest—"traditional" coffee; (2) as a shrub-level crop under a tree-crop overstory (such as oranges or cacao)—"shade" coffee; and (3) as monoculture—"sun" coffee. About forty percent of coffee is grown as sun coffee. Historically, the coffee shrub is derived from an understory plant that requires shade in order to thrive. However, some years ago, a variety was bred that could withstand sun, hence "sun" coffee. This variety lends itself very well to industrial agriculture; in fact, it requires industrial agriculture, that is, use of lots of herbicides, fungicides, and pesticides, which in return require relatively flat land for access by tractors. Such sun coffee plantations are a biological desert.

Based on the experience Charlie and I had investigating coffee growth in Mexico, traditional coffee (grown under natural forest canopy) was indeed a habitat useful to forest-related migrants wintering in the tropics. However, this way of growing coffee has economic drawbacks, mainly that dense shade and high humidity keep production low and make plants susceptible to diseases. In Mexico, the government will provide coffee plants for free to any ejido (communal farm) that wants them. However, the farmers with which we worked found that distance from the market made the cost (fertilizers, fungicides) and work involved in traditional coffee growth uneconomical. Our work demonstrated that the best option for coffee as a cash crop in mountainous country was as a second crop grown under shade provided by another cash crop like oranges. In contrast with traditional coffee plantations, such dual-crop shade coffee plantations, while not exactly a desert biologically, are depauperate in terms of birdlife.

The second experience that I had relevant to Russ's coffee program was our investigation of the Golden-cheeked Warbler. This bird winters in highland pine-oak forest, a habitat that is considered endangered. The reason that it is endangered is that it is being cleared rapidly for pasture and agriculture, and a major part of the "agriculture," in my personal experience, was shade coffee—coffee understory with cacao overstory. We never found a single Golden-cheeked Warbler in such a habitat, and found few of any other bird species.

Thus, when Russ concluded his lecture, I immediately raised my hand to ask what kind of promotion he had in mind. He said that the Smithsonian would promote purchase of shade-grown coffee. I then told him that in my experience, only traditional coffee plantations were good for migrants. Shade-grown coffee, which although technically including traditional coffee, mostly meant a two-crop system unsuitable for migrants. He said that even a two-crop system was better than sun coffee (which is certainly true) and that they would work toward some sort of a verification system to make sure that "shade" meant "traditional."

Well, that did not work. Russ instituted the program, which has been extremely successful. Environment-conscious consumers in the United States are only too happy to pay a premium for shade-grown coffee. However, there is no effective verification program in place to assure that shade-grown coffee is not derived from a two-crop system. The result of this is that the premium paid for shade-grown coffee has the potential to encourage the clearing of pine-oak forest and replacement with the two-crop system on highland slopes formerly considered too steep and uneconomical for clearance, which, of course, is bad for the endangered Golden-cheeked Warbler and its flock cohorts.

Jeepers! Talk about unintended consequences!

Climate Change

> Make thee an ark of gopher wood.
> —Holy Bible, Genesis 6:14 (King James version)

Climate change is real, of course. It has been underway for at least the past 2.4 million years with continuous alteration between warming and cooling cycles. For the past seventeen thousand years, we have been in a warming cycle, especially in the last eight thousand years as glaciers have retreated. The rate of warming has increased within the past century as a result of dramatically increased fossil fuel consumption with accompanying release of carbon dioxide and other gases that serve to trap heat within the earth's atmosphere. We don't know whether such rapid increases would have happened if humans were not here (uncontrolled fire burning forests and prairies and eventually consuming the tundra might have had a comparable effect) or whether similar events occurred in the past twenty cycles. Really, it doesn't matter. Our concern, personally and as a species, is what, if anything, we are going to do about it. And there is the question that we must ask as people concerned with migratory bird conservation—what effect will it have on migrants?

Predictions by experts are that migrants, like most of the rest of the earth's species, will be threatened by climate change. Facts on the ground to date do not support this view. The Texas data discussed in Chapter 8, which are indicative of the findings of many other studies from across the continent, demonstrate that birds as a group, including migrants, adjust rapidly through dispersal to the critical, if indirect, effects of climate change on seasonal timing and distribution of food resources by shifting their ranges accordingly.

These range shifts have, in many cases, caused overlap in distribution of closely related species whose breeding distributions formerly were more or less separate. Biologists, like me, observing these changes realize that they represent a giant, fascinating experiment taking place before our eyes in real time. But experiments have to have hypotheses or they don't really work as a test. Accordingly, I provide the following hypothesis.

Newly overlapping ranges of closely related species will result in one of the following:

- *No change in distribution.* Both species will persist in the newly created environment.
- *Displacement* of the more northern (or higher elevation) member of the species pair by the more southern member.
- *Genetic swamping* (through sexual selection) of the more northern member by the more southern member.

It may be that all three of these results eventually will be seen as a result of the experiment, depending on the species involved. However, I argue that there is evidence that genetic swamping is already underway, and has been for several decades.

The Black-capped and Carolina chickadees comprise a pair of closely related species in which the blackcap has the more northerly or higher elevation distribution as compared with the Carolina. In recent years, hybrids between the two have become more common. In addition, there has been a curious change in the song of the blackcap. Sixty years ago, when I was a child growing up in western New York, the blackcap song was a familiar sound of early spring. Composed of two pure whistled notes, it was described by Elon Edward Eaton in his *Birds of New York* (published in 1910) as "*phe-be.*" Similarly, Roger Tory Peterson, in the first (1934) edition of his *A Field Guide to the Birds*, states in reference to the blackcap, "In the spring he utters a very different call [from 'chickadee'], a two-syllabled whistle: phoe-be." Recordings made in the 1950s also show this call to be composed of two whistled notes, the first at a higher pitch.

After leaving for college, my visits home were brief, and it was not until moving back to the region in 2009 that I had a chance to spend time listening to Black-capped Chickadee songs. I noticed that, although it was still possible to hear the two-note song, most songs seemed to me to be three notes, "*see-dee-dee.*" In 2014, I enlisted Jim Berry, retired president of the Roger Tory Peterson Institute and a skilled birder and recorder of vocalizations, to help me (surprise—he did all the work) in accumulating a library of blackcap recordings. After four years of work, we have found that nearly all vocalizations recorded are of the three-note variety. So what's going on?

One possible explanation is that as climate change progresses, the Carolina Chickadee is better adapted to northern and highland chickadee habitats than its northern counterpart, the Black-capped Chickadee, and that female blackcaps mating with Carolina males produce more offspring. I suggest that the change in blackcap song reflects this influx of Carolina genes into blackcap populations (the Carolina has a four-note song: "*see dee, see doo*") or maybe the effect of female blackcaps choosing to mate with blackcap males that are more "Carolina like."

This story is based mostly on speculation at this point, but I predict that we will see whether or not the theory of southern swamping of northern sister populations is true in fairly short order because there are several of these species pairs. The genetic swamping of the Golden-winged Warbler by the Blue-winged Warbler may be an example.

Such rapid change in species is, of course, entirely inconsistent with the ideas expressed in the paper by Zink and Klicka (discussed in Chapter 9) on the possible relationship between glaciation and speciation events.

According to their work, species like the Carolina and Black-capped chickadees are likely to be separated by at least two million years from the time of their divergence, meaning that none of the twenty or so shifts between glacial advances and retreats over that period has had an effect on their relationship, and certainly not the eight thousand years since forests have reoccupied their temperate ranges.

We shall see.

Conclusion

> SOOTHSAYER. In nature's infinite book of secrecy a little I can read.
> —William Shakespeare, *Anthony and Cleopatra* (Act I, Scene II)

The single most important action that a society can take to protect its migratory birds, or any other aspect of the natural world, is to support unbiased basic research, pay attention to and weigh its outcomes, and assess means of redress when outcomes indicate that they are necessary.

Coda

My purpose in this book has been to provide a comprehensive view of how migratory birds live and where they came from. The Dispersal Theory of migration origin presented in the preceding pages requires a complete rethinking of the migrant way of life. Competition for food occurs mainly on the wintering ground and is mostly or entirely absent from the breeding ground. The breeding season is all about sex, and the breeding territory is no longer seen as a "male dominance" realm (which is the currently accepted concept). It is recognized that females control key aspects of male appearance and behavior with profound implications for breeding system structure (monogamy, polygyny, leks, polyandry, etc.) and rate and directions of migrant evolution. With this recognition in mind, it becomes clear that there are not three seasons in the migrant annual cycle (breeding, migration, and winter) but at least five, each having its own unique requirements for the different ages and sexes of each migrant species: breeding, postbreeding, fall migration, wintering, and spring migration.

The information on which this story has been built was collected by the incredible talent and dedication of hundreds of my fellow field biologists, quick and dead. Part of my purpose in telling this tale in the way that I have is that I wanted to honor that debt by presenting the data in the context of their experience and my own in order to lend some flavor of the field biology enterprise.

At present, field biology as a base from which to construct a theory of bird migration is in disrepute, a perspective led, perhaps, by Robert Martin Zink, distinguished professor and arguably the godfather of the Migratory Syndrome theory. His paper in the *Biological Journal of the Linnean Society* in 2011 provides as clear an exposition of the reasoning behind the phylogenetic approach to understanding the origin and evolution of avian migration as one is likely to find.

Dr. Zink and I go way back. I was instructor for his class in ornithology at the University of Minnesota, and he was my field assistant in Texas and Mexico during the fall and winter of 1973–1974. In fact, he saved John Jr.'s life by stopping me from backing over him with my field vehicle as we were leaving for the evening net check at Welder on October 6, 1973.

After graduation from Minnesota in 1977, Bob went to Berkeley to work with Ned Johnson and received his PhD in 1983. Thereafter our paths diverged. Pug (as my older son, Jay, had dubbed him) developed a stellar career as an eminent molecular systematist and biogeographer while I continued my field studies in bird migration.

In 2002, I received an email from Bob warning me of a commentary he had written on the evolution of migration coming out in the *Journal of Avian Biology*. In this paper, he presented the evidence favoring the Migratory Syndrome hypothesis for the origin of bird migration, and he honored me as the principal proponent of a misguided Dispersal Theory for migration origins (Figure C.1).

He explained that "evidence" in support of this cockamamy idea (my word, not his) was based on a fundamental misunderstanding for how evolution works. His argument is based on the argument presented in the classic 1979 article by Stephen Jay Gould and Richard Lewontin, "The Spandrels of San Marco and the Panglossian Paradigm: A Critique of the Adaptationist Programme," in which they create the mother of all strawmen, dissing the entire discipline of evolutionary biology in general and field biologists in particular (*nota bene*—"adaptationist" is a term applied, in polite company, to field biologists who dabble in evolutionary theory). Their argument was that these deluded saps spent all of their time putting forward harebrained, untestable explanations (as Voltaire's Dr. Pangloss

FIG. C.1. Rappole and collaborators at Gibraltar migration meeting, September 2010. PHOTO BY ROBERT M. ZINK

was apt to do) for every behavior and characteristic observed in an organism, assuming that there had to be some adaptive function in terms of fitness, and proceeding from there to explanations for how the organism evolved. Gould and Lewontin represented this endeavor as absurd, explaining that these flights of fancy were little more than fabrications. They maintained that the proper approach to the study of evolution (as practiced by themselves) was to ignore an organism's traits as of relatively minor importance, perhaps even maladaptive, and focus to the degree possible on the underlying architecture. They used the vaulted ceiling of Venice's St. Mark's Basilica as an example. In attempting to understand the structure, a superficial observer (the field biologist) focuses on the depictions of saints and angels, whereas the cognoscenti (Gould and Lewontin) focus on the arches and spandrels of which the dome actually consists.

Interestingly (to me at least), Benjamin Winger and his associates in their 2014 *Proceedings of the National Academy of Sciences* paper similarly credited me as an author of the dominant (failed) paradigm for the origin of bird migration via dispersal. Which raises an interesting question—why me? Even as solipsistic a person as I am must ask why leaders of the hot, trendy field of migrant evolution as deciphered by molecular genetics should focus their disdain on a minor player—and really on only one publication of that benighted individual, a slim volume published early in his modest career (1995) based largely on his Mexican work and remaindered long ago.

The answer, I think, lies in the fact that the ideas I presented in that book, *The Ecology of Migrant Birds: A Neotropical Perspective,* serve as the perfect foil, the Dr. Pangloss perspective on migration thinking (or maybe Reverend Casaubon). Conclusions reached therein were based almost entirely on speculation derived from field observations ("just so" stories), largely my own and those of fellow field biologists. The contrast with the expensive labs, vast training, impeccable credentials, and splendid publication records for the impressive legion of molecular geneticists involved in generating the phylogenetic data underpinning the Migratory Syndrome hypothesis could hardly be greater.

As I understand the perspective of Dr. Zink and his colleagues, following the analogy of Gould and Lewontin, there is no more basic architecture than DNA—hence the focus on phylogenetic trees to explain migration and migratory histories rather than ecology, which is mere window dressing.

This approach, however, involves a fundamental assumption, namely that "migration" involves a suite of genetic characters unique to that trait, which all migrants must possess. Only by making this assumption can you use phylogenetic trees to trace relationships and draw conclusions regarding whether this migratory species evolved in the temperate zone or the tropics from this or that parental stock. No geneticist has ever identified this suite of characters. The reason, to my mind, is that it does not exist. In essence, it is a plot device necessary to make the whole scheme work. If migration is a form of dispersal requiring no adaptations other than those required for that purpose, then looking for the block of genes that make up a "migration" set, or the switch that turns them on and off for that matter, is as useless as a search for the philosopher's stone.

Be that as it may, the Dispersal Theory certainly rests on shaky ground, basically built on the incorrect assumptions of the Migratory Syndrome as revealed by good old fashioned field biology explained in the preceding chapters of this volume. It is similar in this regard to other theories long scorned for their lack of rigor in the sense of having no known mechanism, such as continental drift, and, like that theory prior to discovery of plate tectonics, it too contains a central mystery, as yet unsolved, namely homing. Despite its apparent ubiquity in the animal world, we know of no mechanism for this behavior. How can young organisms, born in one place, know how to go back to the place from whence their parents originated?

Migratory Syndrome theorists posit that migration is ancient, and that the information regarding how to return to an ancestral home is buried deep in the DNA awaiting only a certain set of environmental circumstances, like bad weather, to trigger it. I agree that migration is ancient, indeed as old as movement. But I do not agree that homing is buried that deeply. Somehow, information on how to return to the place from which the parents originated is passed to the young directly, in a single generation. To me, the migratory switch idea is just another MacGuffin, just like fictitious land bridges connecting southern South America and southern Africa or Australia and Antarctica (before continental drift became accepted)—a concept required by the inconvenient facts on the ground for which there is currently no known explanation. You might just as well posit alien gaslighters shifting stuff around just to drive theoreticians nuts.

The Dispersal Theory presents a completely new understanding of migrants and migration. To summarize:

- Dispersal is a universal phenomenon, common to all forms of life.
- The traits that enhance self-propelled dispersal are the same as those needed for successful migration. These include *Zugdisposition*, *Zugstimmung*, homing, navigation, and orientation.
- Extensive evidence from our Texas work demonstrates the pervasive-

ness (eighty species), power (range change of more than sixty miles), and speed (less than thirty years) of dispersal.

- Examples such as the Bachman's Sparrow demonstrate the ability of dispersal to become migration in a single generation.
- With dispersal as the origin of migration, the expectation is for the more stable environment to be the source of the population with direction of movement to the seasonal environment for breeding.
- As a result, there is an expectation that resident breeding populations will remain in the source area for many species of migrants, an expectation that has been confirmed. Nearly half of all Nearctic migrant species have breeding populations in the same regions where the migratory portions of the population winters.
- Migration depends on availability of superabundant food both along the migration route and at the breeding-ground destination.
- This requirement means that niche strictures are likely to be tightest in the source area from which the population originated and loosest on the breeding ground.
- There are at least five parts to the migrant annual cycle: breeding, postbreeding, fall migration, wintering period, and spring migration. Each part has its own structure and requirements that vary by sex and age, and even for each individual based on the particular environment it encounters.
- Male songbirds arrive several days on average before females. This early arrival (protandry) is not the result of male dominance over females but comes from the fact that the entire structure of the breeding season is based on female choice. Male behavior is structured by sexual selection.
- The breeding period is about maximizing fitness. It is the reason migration exists. Not weather.
- Resources are superabundant during the breeding period, resulting in relaxed niche requirements.
- The purpose of the breeding territory is to demonstrate potential male quality as a partner and provide potential nest sites. Depending on the kinds of foods required to feed the young, it may further provide quality foraging sites.

- Once the young have reached independence, the breeding period is over. The family is no longer the focus. Now each individual is on his or her own, and the behaviors seen during this postbreeding period demonstrate the different motivations for the different age and sex groups—adult male, adult female, young male, young female—including molt, investigation of the vicinity for possible breeding sites in the coming year, and preparation for migration.
- Fall migration results from seasonal change in food resources. Departure timing is built into an annual genetic program for most species of migrants. Weather can play a role, but is not an evolutionary cause of movement. Seasonal climate change is the cause.
- The wintering period generally is the portion of the migrant life cycle where carrying capacity limits populations. Thus, it is during this period that niche strictures are greatest.
- The dichotomy in habitat use seen in many migrants during this period results from lack of appropriate winter habitat. Birds of many migrant species able to locate a space in suitable habitat establish a territory and remain on that piece throughout the winter and return to it in subsequent years. Those unable to locate such a piece, often younger birds, essentially continue in a migratory state moving from one patch of resources to another until they can find a suitable site with resources sufficient to support them. If unable to locate such a site, they continue moving until they die or it is time to head north in spring.
- Timing for spring migration is under genetic control. Weather plays little or no role for most migrants in spring, except perhaps as a modifier of migration speed along the route.
- Spring routes north often differ from routes followed south for the simple reason that conditions along the route in terms of prevailing winds, storm probabilities, and food availability often differ in fall from spring.

Regardless of what we eventually find to be the correct explanation for how migration originates—dispersal, genetic suites, or something else—I hope that the reader will have come to appreciate the importance of field biology, by which I mean, in the broader sense, observation and collection of data directly from the natural world—thus including geologists and astronomers along with botanists and zoologists in the search for new insights taken directly from "out there." This branch of biology has come under significant discredit in recent decades as mere anecdotalism, not really science in any true sense of the word. I understand that perspective. While it used to be that "science" was simply a search for knowledge, Francis Bacon sort of shifted the ground implying that "true science" involved the testing of ideas through hypothesis formulation and experimentation. These aspects of empiricism are extremely important. I do not deny that. But they undermine the value of simple observation.

This obiter dictum opens an additional avenue of consideration. I have mentioned that the shift from gathering of information, often in the form of biological specimens collected in the field, to interpretation of what all these data meant was an important change that occurred in biology during the time of my own training and education in the '60s and '70s. The great theoretical ecologist Robert MacArthur expressed this sentiment quite clearly in the introduction to his book *Geographical Ecology*, published in 1972 (not long before his death of renal cancer at the far too early age of forty-two). He states, "To do science is to search for repeated patterns, not simply to accumulate facts." Thus, succinctly, he opened a new vista of hypothesis formulation, statistical evaluation, and mathematical modeling while relegating those who continued to "accumulate facts" to the dustbin. During the decades preceding this revolution in thought, taxonomists (those concerned with the classification of organisms) were the leaders in helping us to understand how evolution worked, a process based on extensive "accumulation of facts."

This idea changed, even among those who continued to examine the variations in populations who no longer wished to be called taxonomists but "systematists," a term formerly considered synonymous with "taxonomist" now considered to include a new appreciation for higher forms of thought considering the processes of evolution. Robert Selander, whose own work on grackle evolution had depended heavily on collections built by classical taxonomists such as Allan Phillips, had this to say on the subject in his 1971 contribution on systematics to the multivolume work *Avian Biology*: "taxonomy is the only branch of science in which the work of incompetents is preserved." So much for the shoulders of giants.

Neanderthal that I am, I believe that we still have something to learn from those old taxonomists, and from their dusty collections. For me, those hundreds of thousands of dead birds, as well as all of the other organisms similarly preserved, represent vast compendia of genetic information awaiting investigation and interpretation. Many institutions of learning have "disposed" of their collections as useless baggage, expensive to maintain and of no value. I consider their acts comparable to the burning of the Library of Alexandria, a senseless destruction of priceless and irreplaceable information accumulated at great effort and expense.

Now, I think you have the right to ask, "How does this jeremiad relate to migration?" In answer, I will relate two specific situations in which deciphering the genetic information contained in specimens may help in our current understanding of migrant ecology and evolution.

The first is with regard to understanding why many songbird migrants have very high rates of extrapair paternity. Most students of the issue believe that the explanation is technological, which is to say that females have always cheated on their mates but it is only within the past few decades that we have had the ability to detect it. The second part of this statement is undeniably true, but the first part is an assumption. And it is a testable assumption. A remarkable aspect of the extrapair paternity issue is that among closely related tropical species, extrapair paternity is almost nonexistent. Gene Morton and his wife, Bridget Stutchbury, have explained this difference as a result of differences in "synchronicity." They observe that tropical species in general have much longer breeding seasons than migrant relatives to temperate regions. Thus, tropical birds of a given population often are in different stages of the breeding cycle (asynchronous), making it difficult for a female to pick and choose the best quality male with which to mate. In contrast, migrant breeding populations tend to be highly synchronous due to the temperate zone seasonal constraint, with males all in breeding condition at the same time, making it more feasible for a female to select the best one available to sire her offspring.

This explanation is possible, but there are other possibilities, one of which might be that massive destruction of forest habitat in the tropics (greater than ninety percent in many regions) places a much lower premium on age as a factor in survivorship (ability to return to and defend a territory in appropriate habitat) and a much higher premium on chance (stumbling on temporary food superabundances during nonbreeding-season wanderings) allowing much younger males to obtain territory. If this were true, females might accept the territory of a younger male but choose to entrust their genetic posterity to an older, and presumably better adapted, neighbor—a sort of intermediate stage between a territorial breeding system and a lek system. This idea could be tested using egg collections, of which large numbers still remaining from the nineteenth and twentieth centuries. If the explanation that I have given were correct, we would predict that examination for extrapair paternity in clutches of eggs from a century ago would reveal rates much lower than are found today because older males more capable of defending against intrusion by neighbors would make up a greater percentage of territory holders.

A second use of collections relates to speed of genetic change. I have mentioned that molecular systematists assume a standard rate of change based on natural selection acting upon random mutations of two percent per million years. But what if sexual selection plays a role? Would change occur more quickly? I have mentioned that the old taxonomists discovered and reported extraordinarily rapid change in the appearance of populations of very recent origin, like the Bachman's Sparrows and Loggerhead Shrikes from the northeast when compared with their southeastern cousins in a matter of just a few generations. As our ability to read and understand the genetic information in these specimens improves, we may be able to formulate new ways to understand factors that might change the speed of population differentiation in migrants.

The fact that I cannot currently provide additional instances in which the genetic information contained in specimen collections can inform our understanding of migrant evolution stems not from the fact that there aren't more but from ignorance. We are very early in our ability to read and understand the genetic information contained in specimens. A complete Rosetta stone does not yet exist.

I don't want to go all woo-woo on you, but there is a lot that remains mysterious about migration. Maybe not as bad as the Kalahari bushman finding a Coca-Cola bottle in the movie *The Gods Must be Crazy*, but still pretty early in the process. To use an astrophysical metaphor, most people probably think we are about at the Steven Hawking stage of understanding migration. I think we are more at the Newtonian stage or maybe even the Copernican. We know there is a mysterious force, and we can see some of its effects in a crude way. But we are still far from understanding its importance, pervasiveness, or how it works. Molecular systematists will have a hand in that, as will field biologists, and maybe even old-time taxonomists and thinkers from earlier generations.

I will mention one other curious note about ancient ideas regarding migration. The internet is the second-most important development in the history of human civilization, after language. The printing press is a distant third. Consider Homer's theory on crane migration. The great Greek translator Robert Fitzgerald says that the events related in *The Iliad* probably took place about 1300 BC, and that Homer might have been relating them about 850 BC, which means that the crane migration theory likely derived from myths at least three thousand years ago. But consider that Pliny the Elder tells the same story in the section on cranes in his *Natural History*, written from 77 to 79 AD, which doesn't say much for advancement of scientific knowledge regarding bird migration in the intervening millennium. Similarly, papers on swallow hibernation, an idea floated by Aristotle, were still being presented at meetings of the Royal Society, the leading scientific forum in the world at the time, in the late 1800s, long after field data had demonstrated the true situation of southward fall movement. The internet will not allow such ignorance to persist.

The people with whom I have shared my life in field biology have some attributes in common: they are fired by a desire to see what lies beyond the horizon, they want to know all about it, and they don't care what it takes. In addition, they feel a deep sense of privilege for having the opportunity to see Khaos naked, bare of metaphysical raiment.

Most of the great biological theorists, Darwin, Wallace, Ernst Mayr, E. O. Wilson, Robert MacArthur, David Lack, Jared Diamond, and so forth, began their careers as field biologists. Their theories were grounded in what they had seen and read, and their assumptions were clearly laid out. They thought that their theories were the truth, to be sure, but they knew that testing of the assumptions would be the final arbiter of correctness.

Much new theoretical work does not meet this basic standard. Often it is extremely difficult to determine what assumptions have been made (see Appendix 2), and the entire discipline of assumption testing has been relegated to the pejorative field of descriptive biology. How the organism actually lives is considered irrelevant. All that matters is DNA. Eventually this approach will be discredited, and a new crop of field biologists will reexamine the data with fresh eyes and new insights.

No worries.

Appendix 1

Common and Scientific Names of
Bird Species Mentioned in the Text

Presented in alphabetical order by common name following the American Ornithological Society's *Checklist of North and Middle American Birds* and Gill and Wright's *Birds of the World* for all others.

Acadian Flycatcher *Empidonax virescens*
Alder Flycatcher *Empidonax alnorum*
American Golden-Plover *Pluvialis dominica*
American Redstart *Setophaga ruticilla*
American Robin *Turdus migratorius*
American Tree Sparrow *Spizelloides arborea*
Audubon's Oriole *Icterus graduacauda*
Bachman's Sparrow *Peucaea aestivalis*
Bachman's Warbler *Vermivora bachmanii*
Baltimore Oriole *Icterus galbula*
Barred Forest-Falcon *Micrastur ruficollis*
Barred Warbler *Sylvia nisoria*
Bar-tailed Godwit *Limosa lapponica*
Bay-breasted Warbler *Setophaga castanea*
Bewick's Wren *Thryomanes bewickii*
Black-and-white Warbler *Mniotilta varia*
Black-backed Tanager *Stilpnia peruviana*
Blackburnian Warbler *Setophaga fusca*
Black-capped Chickadee *Poecile atricapillus*
Black-chinned Hummingbird *Archilochus alexandri*

Black-crested Titmouse *Baeolophus atricristatus*
Black-faced Antthrush *Formicarius analis*
Blackpoll Warbler *Setophaga striata*
Black-throated Blue Warbler *Setophaga caerulescens*
Black-throated Gray Warbler *Setophaga nigrescens*
Black-throated Green Warbler *Setophaga virens*
Blue-gray Gnatcatcher *Polioptila caerulea*
Blue Grosbeak *Cyanoloxia glaucocaerulea*
Blue-headed Vireo *Vireo solitarius*
Blue Jay *Cyanocitta cristata*
Blue-winged Warbler *Vermivora cyanoptera*
Boat-tailed Grackle *Quiscalus major*
Bobolink *Dolichonyx oryzivorus*
Broad-winged Hawk *Buteo platypterus*
Bronzed Cowbird *Molothrus aeneus*
Brown Creeper *Certhia americana*
Brown-headed Cowbird *Molothrus ater*
Brown Pelican *Pelecanus occidentalis*

Brown Thrasher *Toxostoma rufum*
Buff-bellied Hummingbird *Amazilia yucatanensis*
Buff-breasted Flycatcher *Empidonax fulvifrons*
Canada Warbler *Cardellina canadensis*
Carolina Chickadee *Poecile carolinensis*
Carolina Wren *Thryothorus ludovicianus*
Cassin's Sparrow *Peucaea cassinii*
Cave Swallow *Petrochelidon fulva*
Chestnut-sided Warbler *Setophaga pensylvanica*
Chimney Swift *Chaetura pelagica*
Chipping Sparrow *Spizella passerina*
Chivi Vireo *Vireo olivaceus chivi*
Chuck-will's-widow *Antrostomus carolinensis*
Common Chlorospingus *Chlorospingus flavopectus*
Common Crane *Grus grus*
Common Ostrich *Struthio camelus*
Common Yellowthroat *Geothlypis trichas*
Couch's Kingbird *Tyrannus couchii*
Dark-eyed Junco *Junco hyemalis*
Downy Woodpecker *Dryobates pubescens*
Eastern Bluebird *Sialia sialis*
Eastern Meadowlark *Sturnella magna*
Eastern Phoebe *Sayornis phoebe*
Eastern Towhee *Pipilo erythrophthalmus*
Egyptian Goose *Alopochen aegyptiaca*
Eskimo Curlew *Numenius borealis*
Eurasian Blackcap *Sylvia atricapilla*
Eurasian Widgeon *Mareca penelope*
European Pied Flycatcher *Ficedula hypoleuca*
European Robin *Erithacus rubecula*
Ferruginous Pygmy-Owl *Glaucidium brasilianum*
Field Sparrow *Spizella pusilla*
Golden-cheeked Warbler *Setophaga chrysoparia*
Golden-crowned Kinglet *Regulus satrapa*

Golden Eagle *Aquila chrysaetos*
Golden-winged Warbler *Vermivora chrysoptera*
Grace's Warbler *Setophaga graciae*
Gray Catbird *Dumetella carolinensis*
Gray-cheeked Thrush *Catharus minimus*
Great Blue Heron *Ardea herodias*
Great Egret *Ardea alba*
Greater Prairie-Chicken *Tympanuchus cupido*
Greater Whitethroat *Sylvia communis*
Great Horned Owl *Bubo virginianus*
Great Kiskadee *Pitangus sulphuratus*
Great-tailed Grackle *Quiscalus mexicanus*
Green Jay *Cyanocorax yncas*
Green Kingfisher *Chloroceryle americana*
Groove-billed Ani *Crotophaga sulcirostris*
Hairy Woodpecker *Dryobates villosus*
Hepatic Tanager *Piranga flava*
Hermit Thrush *Catharus guttatus*
Hermit Warbler *Setophaga occidentalis*
Hooded Merganser *Lophodytes cucullatus*
Hooded Warbler *Setophaga citrina*
House Sparrow *Passer domesticus*
House Wren *Troglodytes aedon*
Indigo Bunting *Passerina cyanea*
Kentucky Warbler *Geothlypis formosa*
Kirtland's Warbler *Setophaga kirtlandii*
Least Flycatcher *Empidonax minimus*
Lincoln's Sparrow *Melospiza lincolnii*
Little Blue Heron *Egretta caerulea*
Loggerhead Shrike *Lanius ludovicianus*
Long-billed Thrasher *Toxostoma longirostre*
Louisiana Waterthrush *Parkesia motacilla*
Magnolia Warbler *Setophaga magnolia*
Marsh Wren *Cistothorus palustris*
Mourning Warbler *Geothlypis philadelphia*
Nashville Warbler *Leiothlypis ruficapilla*
Northern Bobwhite Quail *Colinus virginianus*
Northern Cardinal *Cardinalis cardinalis*
Northern Parula *Setophaga americana*

Northern Pintail *Anas acuta*

Northern Waterthrush *Parkesia noveboracensis*

Northern Wheatear *Oenanthe oenanthe*

Olive Sparrow *Arremonops rufivirgatus*

Orange-crowned Warbler *Leiothlypis celata*

Orchard Oriole *Icterus spurius*

Osprey *Pandion haliaetus*

Ovenbird *Seiurus aurocapilla*

Painted Bunting *Passerina ciris*

Painted Redstart *Myioborus pictus*

Palm Warbler *Setophaga palmarum*

Pectoral Sandpiper *Calidris melanotos*

Peregrine Falcon *Falco peregrinus*

Pileated Woodpecker *Dryocopus pileatus*

Pine Warbler *Setophaga pinus*

Prairie Warbler *Setophaga discolor*

Prothonotary Warbler *Protonotaria citrea*

Purple Martin *Progne subis*

Red-eyed Vireo *Vireo olivaceus*

Red-headed Woodpecker *Melanerpes erythrocephalus*

Red Knot *Calidris canutus*

Red-winged Blackbird *Agelaius phoeniceus*

Resplendent Quetzal *Pharomachrus mocinno*

Rock Pigeon *Columba livia*

Rose-breasted Grosbeak *Pheucticus ludovicianus*

Ruby-throated Hummingbird *Archilochus colubris*

Scarlet Tanager *Piranga olivacea*

Slate-colored Solitaire *Myadestes unicolor*

Snowy Egret *Egretta thula*

Song Sparrow *Melospiza melodia*

Spot-crowned Woodcreeper *Lepidocolaptes affinis*

Spotted Antbird *Hylophylax naevioides*

Stonechat *Saxicola*

Sulphur-rumped Flycatcher *Myiobius sulphureipygius*

Summer Tanager *Piranga rubra*

Swainson's Thrush *Catharus ustulatus*

Swainson's Warbler *Limnothlypis swainsonii*

Tennessee Warbler *Leiothlypis peregrina*

Townsend's Warbler *Setophaga townsendi*

Tropical Parula *Setophaga pitiayumi*

Tufted Titmouse *Baeolophus bicolor*

Unicolored Jay *Aphelocoma unicolor*

Veery *Catharus fuscescens*

Western Wood-Pewee *Contopus sordidulus*

White-breasted Nuthatch *Sitta carolinensis*

White-breasted Wood-Wren *Henicorhina leucosticta*

White-eyed Vireo *Vireo griseus*

White-necked Thrush *Turdus albicollis*

White-throated Bee-eater *Merops albicollis*

White-throated Sparrow *Zonotrichia albicollis*

White-tipped Dove *Leptotila verreauxi*

White-winged Dove *Zenaida asiatica*

Whooping Crane *Grus americanus*

Willow Flycatcher *Empidonax traillii*

Willow Warbler *Phylloscopus trochilus*

Wilson's Warbler *Cardellina pusilla*

Winter Wren *Troglodytes hiemalis*

Wood Thrush *Hylocichla mustelina*

Worm-eating Warbler *Helmitheros vermivorum*

Yellow-bellied Flycatcher *Empidonax flaviventris*

Yellow-billed Cuckoo *Coccyzus americanus*

Yellow-breasted Chat *Icteria virens*

Yellow-throated Vireo *Vireo flavifrons*

Yellow-throated Warbler *Setophaga dominica*

Yellow Warbler *Setophaga petechia*

Appendix 2

A Critical Examination of the Assumptions in "Temperate Origins of Long-Distance Seasonal Migration in New World Songbirds" by Benjamin M. Winger, F. Keith Barker, and Richard H. Ree

The critique presented below represents a departure from my avowed intent to eschew the normal procedures for presentation of a technical paper in my discipline in favor of a more reader-friendly narrative. My reason for departing in this way is that I consider the paper by Dr. Winger and his colleagues to be an extremely important reformulation of the Northern Home Theory of migrant evolution from a molecular genetics perspective. As such, it warrants a detailed response according to kosher scientific procedure and jargon, as would have happened in *The Avian Migrant* had the paper not come out too late (2014) for treatment in that book.

Reviewers have pointed out that this appendix will be heavy going for my intended audience. No doubt they are right. The models presented are complicated, and the assumptions on which they are built are not clearly put forward by the authors. Discerning these assumptions from the text is no small chore, involving intensive study. In addition, analysis of their validity requires some detailed knowledge of, and reference to, migrant ecology. Nevertheless, I believe that close study of the argument will enable the interested reader to understand the conclusions.

The article by Winger et al. (2014) raises important questions regarding phylogenies and their use in determination of how, when, and where migration and diversification occurred in a large group of New World songbirds, the Emberizoidea (sparrows, blackbirds, orioles, wood warblers, tanagers, and their close relatives). In this paper, the authors propose to answer the following question: Did migration in the New World nine-primaried oscines evolve "via shifts of the breeding range out of the tropics, driven by increased reproductive success and reduced competition in temperate regions" or "when species resident year-round in temperate latitudes shift their winter ranges to lower latitudes [i.e., the tropics] to increase survival during the harsh and resource-depleted temperate winters" (Winger et al. 2014:1)?

Using the phylogenetic tree developed by Barker et al. (2015) and a clever series of decision matrices and statistical treatments, the authors reconstruct ancestral distributions along the entire lineage of the New World Emberizoidea, dating back fourteen million years. Based on this reconstruction, they reach sweeping conclusions regarding the origin of migration and diversification of this group in the New World. These conclusions, if correct, have profound implications regarding our understanding of avian population dynamics, ecology, migration, and evolution. In addition, and perhaps even more significantly, the approach taken to reach these conclusions is revolutionary. The only information used in the analysis is the current range for the 823 species in the group and DNA data indicative of the phylogenetic lineage branchings that purport to connect these modern species all the way back fourteen million years to a hypothesized protoemberizoid.

The purpose of this appendix is to examine critically the information, procedures, and assumptions on which the conclusions in Winger et al. are based.

Methods Summary

The methods used by Winger et al. to reach their conclusions involve no ecological arguments or analyses. Instead, the authors use data on the cur-

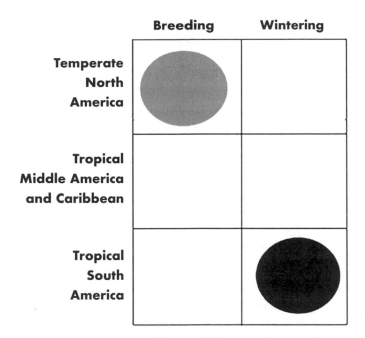

FIG. A2.1. Range summary "domino" for a neotropical migrant, e.g., the Blackpoll Warbler, that breeds (*gray circle*) in temperate (or boreal) North America and winters (*black circle*) in tropical South America.

rent range (breeding and wintering), as derived from Ridgely et al. (2003), Poole (1992–2010), and del Hoyo et al. (1992–2014). They use these data to construct crude summaries called "dominos" (because of their similarity to the game piece) (Figure A2.1). These summaries show presence or absence as breeding (column 1) or wintering (column 2) populations for all New World emberizoids in three regions: temperate North America (row 1); tropical Middle America and the Caribbean (row 2); or tropical South America (row 3).

The range summaries are used along with phylogenetic data to construct a model for inferring the biogeographic history of migratory lineages. The emberizoid phylogeny used in the model was constructed by Barker et al. (2015) using DNA evidence to trace back each lineage to the point at which speciation might have occurred (cladogenesis), beginning with the

TABLE A2.1.

"Rules" governing transitions from ancestral range to daughter ranges in the emberizoid lineage (Winger et al. 2014 Supp.)

1. Both breeding and winter range [of the two daughter species] can show expansion or contraction separately or in unison into neighboring regions [i.e., temperate North America, Middle America and the Caribbean, and tropical South America] relative to the range of their immediate ancestor.
2. The new breeding range [of the two daughter species] cannot expand or contract to result in breeding-range occupancy into regions south of the winter range [of the immediate ancestor].
3. The new winter range [of a daughter species] cannot occupy a region north of the breeding range [of the ancestor].
4. Breeding-range expansion [for a daughter species] into a region south of the [ancestral] breeding range must be accompanied by comparable expansion of the winter range.
5. Winter range expansion [for a daughter species] into a region north of the [ancestral] winter range must be accompanied by a comparable expansion of the breeding range.
6. Breeding range [for a daughter species] can expand into, or contract from, a region north of the [ancestral] breeding range without change in the winter range.
7. Winter range [for a daughter species] can expand into, or contract from, a region south of the [ancestral] breeding range without change in the breeding range [of the daughter species].
8. Neither breeding nor wintering range [for a daughter species] can jump a region [as compared with the ancestral species], e.g., an ancestral breeding range in South America cannot give rise to a daughter species' breeding range in temperate North America.
9. There is, however, an exception to rule 8: an ancestral migratory species with a breeding range in temperate North America can have a daughter species that breeds in tropical South America.
10. During speciation, the ancestral range can be inherited by daughter species or fragmented, but the entire ancestral range must be cumulatively present in the daughter species.
11. Migratory daughter species cannot be derived from a nonmigratory ancestor.

Note: The list of "rules" is derived from the "Supporting Information" portion of the paper (Winger et al. 2014 Supp.). They are not cited here exactly as written, and they include some explanatory notes not provided in the text.

current 823 species of emberizoids and following each branch back from there fourteen million years to the assumed first New World member.

In essence, the model is a decision matrix in which the decisions are referred to as "transitions." These transitions involve the change for the daughter species from the range of the ancestral species to the new range of the derivative species. Winger et al. provide several "rules" for these transitions (Table A2.1). These "rules" allow them to conclude what the range of the ancestor was from the ranges of the two daughter species, which in turn provides the basis for their models and statistical treatments.

Using these "rules," along with the current ranges of all New World emberizoids, i.e., all of the currently existent daughter species, the authors reconstruct the ranges of all species ancestral to these living daughter species all the way back to the first ancestor, an assumed fourteen million years ago (Ma), at least at the level of their occupancy of the three defined regions.

The authors use these reconstructions to determine whether or not ranges of daughter species tended to shrink (occupy fewer regions) or split (occupy different parts of the same region) when compared with ancestors. To provide a metric of this change, they assigned a "range separation" value to each emberizoid species throughout the entire phylogenetic tree, the current ranges being used along with the rules to derive the ancestral ranges all the way back to the first ancestor. This range separation metric was calculated as the number of different regions occupied in the breeding season but not in winter plus the number of regions occupied in the winter but not in the breeding season plus the number of regions separating each. Thus highest range separation would be a value of 4, which would occur for a species that breeds only in temperate North America (+1) and winters only in tropical South America (+1), with both breeding and wintering range being separated by a region (Middle America and the Caribbean) (+2). Lowest range separation value (0) would be for a species that breeds and winters in the same region. Each emberizoid species in the entire lineage dating back to the assumed first member is thus assigned a range separation value. This value is compared with that of its putative ancestor by subtracting the range separation value for a daughter from the range separation value for an ancestor in order to calculate whether ranges in emberizoids tend to shrink or remain similar from ancestor to daughter. For instance, if an ancestral species that breeds in temperate North America and winters in tropical South America (range separation value of 4) has a daughter species whose breeding and wintering range is tropical South America (range separation value 0), then the difference is 4. Thus, transitions with a positive value show decrease in range from ancestor to daughter, while transitions with a negative value show increases.

As in the case of domino construction described above, determination of the range of the ancestor species is critical to this analysis. To address this issue, the authors generated a "posterior probability distribution of ancestral states" using the following procedures (Winger et al. 2014 Supp.:2):

> For each posterior sample of μ and Q from the Markov Chain Monte Carlo (MCMC) runs, the conditional likelihoods of ancestral states given the observed dominos at descendant leaf nodes were computed for each internal node using the pruning algorithm of (17) [Felsenstein 1981]. These likelihoods were used to compute the maximum-likelihood set of ancestral states for μ and Q, first at the root node (i.e., the state with highest conditional likelihood), and then by a recursive preorder traversal of the phylogeny, selecting the state with highest likelihood given the domino at the start of the node's ancestral (subtending) branch.

In other words, in order to figure out the range of the ancestor, they looked at the range of the daughters and, based on a set of assumptions, calculated a probability as to which of the daughter ranges most resembled that of the ancestor's, and they assigned the range of that daughter to the ancestor. The assumptions used for this calculation are not stated, but presumably they are the same "rules" as those used in the range transition analysis.

Evaluation of Results and Conclusions

After performing the treatments described above on the range and ancestry data, Winger et al. then calculate the number of transitions in which the range separation value increased from ancestor to daughter or decreased. They find that in the vast majority the number decreased, i.e., daughter species tended to occupy fewer regions for breeding and wintering than did ancestors.

Based on these and related analyses, the authors reach the following conclusions:

1. The first emberizoid colonist to the New World arrived via dispersal from east Asia across Beringia into the northern regions of temperate North America.
2. This first colonist was a temperate zone resident.
3. Over a long period, seasonal deterioration in weather pushed this first temperate zone resident southward during temperate winter, eventually resulting in the evolution of long-distance migration.
4. Some members of this long-distance migrant population wintered as far south as Middle America, the Caribbean, and South America.
5. Some of the wintering members of this population remained to breed in their wintering grounds, establishing resident populations in the tropics. This behavior is referred to as "migratory drop-off" by the authors.
6. Subsequent diversification of the group in the tropics of Middle and South America occurred in three ways: "fragmentation of widespread ranges, expansion of resident ranges of nonmigratory populations from temperate to tropical latitudes, and the expansion of a migratory species' breeding range into a tropical area already occupied by their winter range" (migratory drop-off) (Winger et al. 2014).
7. They emphasize that the most important of these sources involved early evolution of migration in the first colonist for temperate North America followed by migratory drop-off in the tropics, resulting in tropical diversification. The authors point to the tanager family, Thraupidae, as the premier example of this process in action. According to Barker et al. (2015), this family arose roughly 13 Ma and is restricted in distribution entirely to the New World tropics of Mid-

dle and South America, at least as it is currently recognized by Barker et al. This family is the largest group of New World emberizoids. It consists of 374 species. They are pantropical in the New World, but their greatest diversity occurs in the northern Andes. Most of the other emberizoid families also have their centers of diversity located in the tropics. As hypothesized for Thraupidae, Winger et al. propose that these families also arose through migratory drop-off of temperate zone emberizoids followed by in situ diversification in the tropics.

8. Long-distance migration between North America and the neotropics has evolved many times in this group, principally if not entirely due to shifts of winter ranges from North America into the tropics as opposed to breeding-range shifts from tropical latitudes into temperate regions.

9. Shifts of the winter range imposed by seasonal climate deterioration are the main drivers of the evolution of long-distance migration throughout Emberizoidea and, probably, most other migratory species as well.

10. Long-distance migration is the primary method through which temperate zone emberizoids established resident breeding populations in the tropics. This occurred through "migratory drop-off," i.e., when wintering birds of temperate breeding populations remained on the wintering ground to breed. Subsequent diversification within the tropics occurred via the normal processes of fragmentation of widespread ranges.

11. Expansion of resident ranges of nonmigratory populations from temperate to tropical latitudes also occurred in some instances, but at a much lower frequency than migratory drop-off.

Summarizing their hypotheses, Winger et al. state, "*we found that long-distance migration primarily evolved through evolutionary shifts of geographic range south for the winter out of North America, as opposed to north for the summer.*" Note that they do not couch their findings with any degree of doubt. It is as though they are reporting the testing of a crop treatment, like, "we found that addition of nitrogen increased yield by eighteen percent." Yet there is a very great deal of doubt, as discussed below.

Evaluation of Assumptions Required

Most of the hundreds of assumptions made by the authors to reach their conclusions are not mentioned. Those assumptions that are mentioned are called "rules," specifically, "*the rules governing transitions between dominos*," which refers to the calculated transition from the range of the ancestor to the ranges of the daughters. The reason these assumptions are referred to as "rules" is presumably because the logic of the analysis *requires* that they *always* be correct. Deviations from the "rules" derail the line of reasoning that leads back to the range of the first ancestor.

A few of the "rules" (Table A2.1) are logical and unexceptionable. Several, however, are not based on known principles of ecology, natural history, or biogeography, as discussed elsewhere in this volume. Some examples are discussed below.

"Rule" 4. Breeding-range expansion for a daughter species into a region south of the ancestral breeding range must be accompanied by comparable expansion of the winter range. Hundreds of exceptions to this "rule" are known to occur. Many species of birds that are resident in the South American, Asian, and African tropics appear to have given rise to closely related migratory populations that breed in temperate regions south of the tropical breeding and wintering areas of their presumptive ancestors but winter north of their breeding areas in the tropics (McClure 1974, Jones 1999, Chesser and Levey 1998, Chesser 2005), including at least one species of the emberizoid family, Thraupidae, the Black-backed Tanager (*Stilpnia peruviana*) (Hilty 2011:229). The authors acknowledge that this is the case, but they claim that no examples are known for the three regions defined (temperate North America, Middle America, and tropical South America). So far as I am aware, this assertion is correct. However, by stating categorically that an apparently common "transition" cannot occur, the authors limit consideration of the evolution of migration to a specific set of assumed circumstances (i.e., a north temperate zone ancestor for every tropical daughter species),

and in so doing, they dictate a limit regarding understanding of how migration evolves.

"Rule" 8. *Neither breeding nor wintering range for a daughter species can jump a region as compared with the ancestral species, e.g., an ancestral breeding range in South America cannot give rise to a daughter species' breeding range in temperate North America.* There is a significant body of evidence available to question the validity of this assumption. Closest relatives for many species of migrants from tropical South America to the temperate region of North America are tropical South American residents. For instance, the Red-eyed Vireo's (*Vireo olivaceus olivaceus*) closest relative (considered by some authors to be a separate species) is the Chivi Vireo (*Vireo olivaceus chivi*), a resident or partial migrant of the Amazon basin. There are many similar examples from tropical Africa and Asia as well. One could argue, as the authors have, that all of the tropical relatives of long-distance migrants to the temperate zone are the result of "migratory drop-off." This argument seems highly unlikely given the abundant ecological and taxonomic evidence to the contrary (Rappole 2013 and this volume), but it is not impossible. However, the fact that it is not impossible does not allow its qualification for "rule" status.

"Rule" 9. *There is, however, an exception to rule 8: an ancestral migratory species with a breeding range in temperate North America can have a daughter species that breeds in tropical South America.* This assumption is the "migratory drop-off" rule from which, the authors believe, most South American emberizoids are derived, either directly or indirectly. Taken in combination with "rule" 8, it allows the authors to reconstruct ancestral ranges entirely consistent with the drop-off hypothesis. However, if there are exceptions to "rule" 8 in which tropical residents give rise to long-distance migrants, "rule" 9 is circular and invalid. Much of the current volume has been devoted to explaining how the ecology and natural history of migrants support derivation from tropical residents. Winger et al. do not include ignoring of these data as assumptions. They argue that the phylogenetic tree plus the rules means it simply didn't happen that way.

"Rule" 10. *During speciation, the ancestral range can be inherited by daughter species or fragmented, but the entire ancestral range must be cumu-*

latively present in the daughter species. This assumption allows the authors to assume that the most complex range (i.e., combination of the two daughter ranges) is always the ancestral range. This assumption is absolutely critical to construction of ancestral ranges, especially the farther back you go in time along the lineage. Following this rule allows the authors to go all the way back to the first member. However, it can only be valid if "rule" 11 is correct. If there are exceptions to "rule" 11, then "rule" 10 is circular and invalid.

"*Rule" 11. Migratory daughter species cannot be derived from a non-migratory ancestor.* This assumption is the key on which all of the other "rules," and indeed the entire argument of the paper, hinges. If it is correct, then the key parts of the argument hold. If there are exceptions, then the entire methodology is circular and invalid. This assumption has to be true in order for the authors to perform the analyses that prove it is true. Of course, it *may* be true, but that is not enough. As the authors note, there are several migration theorists who have argued that most temperate zone migrants derive from tropical zone residents. Perhaps arguments of these "southern home" theorists do not achieve the level of a "paradigm" as the authors claim, but certainly they have published a considerable amount of field data in support of their thesis (Dixon 1897, Mayr and Meise 1930, Williams 1958, Rappole et al. 1983, 2003, Cox 1985, Levey and Stiles 1992, Rappole and Tipton 1992, Rappole 1995, 2013, Safriel 1995, Rappole and Jones 2002, Helbig 2003, and this volume). Logical process requires that Winger et al. answer these arguments before they dismiss them. Simply mentioning them in disparaging terms is not sufficient, especially when the procedures, models, and statistical treatments on which the conclusions in Winger et al. are based are circular.

Conclusions

The strength of the approach taken in Winger et al. is that, if the methods and reasoning are sound, then the conclusions are irrefutable. Thus, they do not couch their conclusions in the usual cautious manner used in hypothetico-deductive reasoning, as in "our results indicate . . ." Instead, they state them more in the form of a mathematical proof ending with an implied QED, as in "we found that long-distance migration primarily evolved through evolutionary shifts of geographic range south for the winter out of North America, as opposed to north for the summer" (Winger et al. 2014:1). I have presented above an analysis of the assumptions, which they call "rules," to which they admit. However, their treatment actually includes hundreds of additional assumptions that they do not address. These assumptions fall into five major categories, as discussed in the text in Chapter 10.

To summarize, then, the entire support for origin of migration via the Migratory Syndrome in tanagers and their relatives according to Winger et al. derives from two forms of data, phylogenetic trees and current ranges. As outlined in the above critique, and documented in this book, this conclusion is absurd. But what is the alternative? Several other explanations are possible for the colonization by emberizoids of the New World, their subsequent invasion and diversification in the tropics, and evolution of migration in the various members of the group. I discuss some of these below under the headings of the major conclusions of Winger et al.: "Entry into the New World," "Colonization and Diversification in the Neotropics," and "Evolution of Migration."

Entry into the New World. The authors assume that the first emberizoid colonists to the New World arrived via dispersal across Beringia about 14 Ma. Thus this progenitor of New World Emberizoidea is assumed to be a north temperate zone resident. These assumptions derive from the phylogenetic work of Barker et al. (2015). Figure 1 in Barker et al. shows the entire phylogeny of the New World Emberizoidea, and it provides the assumed region of origin for each major branch of the phylogenetic tree. At the base of the tree, the first New World member of Emberizoidea, from which all other emberizoids are assumed to have been derived, is shown as a resident of the north temperate zone of North America. Barker et al. do not know that this is true. They have no fossils documenting the presence of this protoemberizoid in Beringia or northern North America, and even if they did, one cannot tell from a fossil whether it is a resident of the temperate zone or a migrant that breeds in the temperate zone, winters in the tropics, and perhaps evolved in the tropics. Nevertheless, the assumption is logical. Beringia served as a connection between the Eurasian and North American land masses of varying width north to south for most of the past one hundred million years (Sewall et al. 2007, Fiorillo 2008). As such, it served as the main pathway for New World colonization by many different groups of animals and plants for nearly sixty million years (Pigg et al. 2004, Vila et al. 2011, Guo et al. 2012).

However, it is not the only possible pathway for invasion, at least for birds. In fact, the first emberizoid to colonize the New World could have flown there. And, if that was the mode of entry, then the most likely place of entry was tropical South America, since the very first branching of the phylogenetic tree constructed by Barker et al. produces South American daughter species. Indeed, over half of all emberizoids are South American residents. But is such a founding flight possible? As Barker et al. note, there are other examples of long distance colonizations by emberizoids. It was emberizoids that colonized the Galapagos, 1,800 km from the nearest land, and emberizoids that colonized Tristan da Cunha Island in the South Atlantic, 2,900 km from southern Africa. Also, as if these historical examples were not enough, there are current examples. For instance, Brosset (1968) reported that a small population of two New World migrants, the Red-eyed Vireo (*Vireo olivaceus*) and the Summer Tanager (*Piranga rubra*) (an emberizoid), winter regularly in Gabon in equatorial Africa, 6,000 km outside their normal winter range. Clearly, these examples demonstrate that the first New World emberizoid could have entered tropical South America directly from tropical Africa rather than indirectly via Beringia.

Thus, a tropical South American entry for Emberizoidea into the New World from Africa is possible. Of course, if that were the mode of entry, then the authors would have to explain how these birds moved northward across ocean to Middle America and the Caribbean—no land connection was present until about 4.6 Ma (Haug and Tiedemann 1998)—and also how a number of long-distance migrants to the temperate zone evolved in the absence of significant climate change at point of origin (i.e., the tropics).

Colonization and Diversification in the Neotropics. Let us assume, however, for the sake of argument, that the first emberizoid entered via the

Bering continental connection as Barker et al. have assumed. Then how did this first temperate zone resident colonist produce a major lineage of South American species very early in the evolution of the group? Winger et al. assume that this protoemberizoid developed migration to South America followed by migratory drop-off. However, the protoemberizoid could first have colonized the New World by flight, directly from Africa, as described above. A third possibility is that seemingly followed by the *Polyommatus* butterflies (the Nabokovian Hypothesis, Vila et al. 2011—yes, *that* Nabokov, of *Lolita* fame), which invaded the New World via Beringia at least five times over the past eleven million years, expanding southward each time, and entering South America during periods when that continent was separated from North America by at least 100 km of open ocean by the Atrato Seaway (Kirby et al. 2008). A similar route for invasion and colonization of the Nearctic and neotropics apparently was followed by natricine snakes, although this group has not yet invaded South America (Guo et al. 2012). This group evidently evolved in the Old World tropics of Asia during the Eocene, dispersing into Africa in the Oligocene (28 Ma) and into the Palearctic and New World across Beringia 27 Ma.

Whether it was migratory drop-off or dispersal from the New World temperate zone or Old World tropics, the new arrival in tropical South America would confront the same problem presented to any bird attempting to enter from the outside region, namely, "Moreau's paradox" (Moreau 1972). How can an individual from outside survive, and, in the case of a migratory drop-off scenario, breed, in a complex, highly evolved avian community like that which existed in South America 13–14 Ma? As discussed in Chapter 10, there are three possible answers to the paradox: (1) by exploiting excess or superabundant resources, (2) by locating empty niches, or (3) by not being from the outside, i.e., essentially returning to the place of its origin.

Since the first colonist cannot be returning to its origin, the third possibility is denied it, and since it stays to breed, excess resources also are not a likely explanation. Therefore the second explanation seems most likely, i.e., the first colonist was able to fill an empty niche. There had to be a new environment available to this new colonist, otherwise highly evolved resident avifauna would have outcompeted them. It may be that Andean orogeny presented just such an environment (Gutierrez et al. 2013).

Existence of such a newly created environment, one to which lowland forest residents were not well adapted to exploit, could explain how the new arrivals were able to colonize South America 10 or 11 Ma, whether they came from the Nearctic or the African tropics. However, this hypothesis cannot explain how emberizoids from the Nearctic might have invaded neotropical environments more recently, as both Winger et al. and Barker et al. suggest (through migratory drop-off). So far as we know, there have been no vast areas of empty niches to invade for several million years. So how did they do it? The temperate zone origin hypothesis of migratory drop-off does not address this question. Winger et al. simply state that these temperate emberizoid colonists just did it, as proven by the transition maps. All previous theorists recognize that this ecological question has to be explained, and so far, only three explanations have been provided: superabundant resources, empty niches, or returning migrants of tropical origin.

Evolution of Migration. Winger et al. present the following theory for the evolution of migration.

1. The first emberizoid to enter the New World was a temperate zone resident.
2. These temperate zone resident populations are subjected to increasingly severe weather seasonally during the winter period.
3. Over a long period of time, natural selection favors evolution of a suite of characters that allow some members of the population to temporarily escape the weather by migrating south for the winter.
4. These birds return in summer to breed.

They contrast their theory to a single theory of southern home origin in which birds that breed in the Nearctic and migrate to the neotropics for the winter originated in the neotropics. This approach tends to obscure the true differences between their theory and others. The real differences are the way in which the influences of resources and time are understood. In the Winger et al. theory, movement is a necessity, i.e., lowered survival due to lack of resources drives evolution of migration. Responding to this necessity takes evolutionary time to develop. The birds can't just leave when they need to as the resources decline. They have to wait generations for proper adaptations to evolve. This concept is quite different from the theory proposed in this volume in which movement is an opportunity, and, because it is an opportunity, it is always a possibility for any dispersing individual. All dispersing individuals of all species already possess the basic equipment to capitalize on the opportunity, for example, the ability to fly, to fatten, and to find their way. According to this theory, migration is a form of dispersal. Contrary to the claims of Winger et al., the theory does not necessarily envision a southern home origin. Rather, what the theory proposes is that dispersal, usually by young of the year, is random in terms of direction. The dispersers move away from competition with conspecifics in search of new areas with suitable habitat and available resources to colonize. If they find it, they settle. If the habitat is aseasonal, they stay. If it is seasonal, they stay until resources deteriorate, and then return to where they came from. Thus, migration is predicted to develop from the less seasonal environment, where the colonizing birds originated from a resident population, to a more seasonal environment, where the colonists go to breed. In the Northern Hemisphere, this seasonal movement will more often be northward in spring toward areas where resources are seasonally superabundant and southward in fall. In the Southern Hemisphere, the pattern will be reversed. In places where the seasonal environment is a mountain, the movement will be upslope in spring and downslope in fall; in areas where rainfall rather than temperature is the driver of resource availability, the movement will follow seasonal rainfall patterns, which could be north, south, east, or west of the less seasonal environment—as is the case in sub-Saharan Africa and much of Australia.

Of course, the "migration as dispersal" theory is hypothetical. I provide extensive ecological evidence for its existence throughout this volume, but I do not prove it. Nevertheless, I think that the argument is well supported by ecological data on migrants from across the world. The same cannot be said for the Winger et al. hypothesis, which rests solely on a specious statistical treatment of phylogenetic data.

Appendix 3

Notation Corrections for Alan Pine's
Multiple Carrying Capacity Equations from
"Age-Structured Periodic Breeders"
by Alan S. Pine in *The Avian Migrant:*
The Biology of Bird Migration
by John H. Rappole
(New York: Columbia University Press, 2013)

This appendix is included because of typesetting errors that occurred in Appendix B of *The Avian Migrant* without the opportunity for the authors to correct them. In Appendix A of that volume, Dr. Pine described mathematical generalizations of the classical Verhulst logistic expression governing the population dynamics of a species in a single habitat with a dominant carrying capacity. Therein, the role of distinct density-dependent birth and death factors in multiple periodic (i.e., seasonal) habitats was discussed for application to migratory species. Both immediate and delayed response to limiting mechanisms was considered, with the latter possibly leading to chaotic patterns in the event of excessive birth rates. Threshold effects and metamorphic (non-overlapping generations) species were also described. In Appendix B, age-dependent reproduction and mortality were included for a more realistic description of the dynamics of important migrating or seasonal species. This treatment necessitated vector and matrix mathematics with their accompanying more complex notation and symbols, which was the principal reason for the typographical errors corrected in the following Erratum.

APPENDIX B ERRATA: "Age-Structured Periodic Breeders" by Alan S. Pine in *The Avian Migrant: The Biology of Bird Migration* by John H. Rappole, Columbia University Press, New York, 2013, pp. 330–344.

PAGE	LINE	CHANGE	TO
332	11	m	M
332	21	$x_{M+1,n}$	$x_{m+1,n}$
332	22	$x_{M+1,n} = \mathrm{x}_{m,n}[\cdots]$	$x_{m+1,n} = x_{m,n}[\cdots]$
335	1	B_{C0}	b_{C0}
337	9	$\mathrm{Y}_{m+1,n} = [:] = \mathrm{A}_{m,n} \bullet \mathrm{Y}_{m,n}$	$\mathbf{Y}_{m+1,n} = [:] = \mathbf{A}_{m,n} \bullet \mathbf{Y}_{m,n}$
337	14	$\mathrm{Y}_{1,n+1} = [:] = \mathrm{A}_{M,n} \bullet \mathrm{Y}_{M,n}$	$\mathbf{Y}_{1,n+1} = [:] = \mathbf{A}_{M,n} \bullet \mathbf{Y}_{M,n}$
338	2	$\mathrm{Y}_{1,n+1} = \widehat{\mathrm{A}}_n \bullet \mathrm{Y}_{1,n}$	$\mathbf{Y}_{1,n+1} = \widehat{\mathbf{A}}_n \bullet \mathbf{Y}_{1,n}$
338	3	$\widehat{\mathrm{A}}_n \triangleq \mathrm{A}_{M,n} \bullet \displaystyle\prod_{m=1}^{M-1} \mathrm{A}_{m,n}$	$\widehat{\mathbf{A}}_n \triangleq \mathbf{A}_{M,n} \bullet \displaystyle\prod_{m=1}^{M-1} \mathbf{A}_{m,n}$
338	14	$\widehat{\mathrm{A}}_n =$	$\widehat{\mathbf{A}}_n =$
338	23	$\widehat{\mathrm{A}}_n$	$\widehat{\mathbf{A}}_n$
338	27	$\left|\widehat{\mathrm{A}}_n - \lambda_n \mathbf{1}\right| = 0$	$\left|\widehat{\mathbf{A}}_n - \lambda_n \mathbf{1}\right| = 0$
339	22	$\widehat{\mathrm{A}}_n$	$\widehat{\mathbf{A}}_n$
339	30	$\widehat{\mathrm{A}}_0$	$\widehat{\mathbf{A}}_0$
339	33	$Y_{1,0}$	$\mathbf{Y}_{1,0}$
339	39	$x_{M+1,n}$	$x_{m+1,n}$
339	40	$A_{m,0}$	$\mathbf{A}_{m,0}$
339	41	$Y_{1,0}$	$\mathbf{Y}_{1,0}$
339	41	$A_{m,n} = A_{m,0}$	$\mathbf{A}_{m,n} = \mathbf{A}_{m,0}$
340	12	x_m	x_M
340	15	x_m	x_M
340	15	y_m	y_M
341	3	$\widehat{\mathrm{A}}_n$	$\widehat{\mathbf{A}}_n$
342	5	$\widetilde{\mathrm{A}}_n$	$\widetilde{\mathbf{A}}_n$
342	9	$\lambda_{k-1}\beta_1$	$\lambda^{k-1}\beta_1$
343	22	$x_{M+1,n}$	$x_{m+1,n}$
344	29	$\widehat{\mathrm{A}}_n$	$\widehat{\mathbf{A}}_n$

Annotated Bibliography

The citations provided below are heavily skewed in favor of my own views regarding the critical ideas related to bird migration, not only with respect to the publications included but, in particular, in my comments wherein I indicate why the publication is recommended for my readers.

American Ornithological Society. *Birds of the World.*
 https://birdsoftheworld.org/ (accessed March 24, 2020).
 This publication represents the combination of *The Birds of North America*, a successor to Arthur Cleveland Bent's *Life Histories of North American Birds*, and the *Handbook of the Birds of the World*. It contains life histories for all of the world's bird species, as understood in 2020. It is the gold standard concerning the world ornithological community's understanding of avian life history. From the perspective of our book, the importance is the continued expression of a lack of understanding of the nonbreeding portion of the life cycle in general, and a complete absence of recognition of a "postbreeding period" for those migrants in which it is known to occur, as well as the lack of understanding of the fact that spring and fall migration are quite different in many (most?) species of migrants (often involving completely different routes).
———. *Checklist of North and Middle American Birds.*
 http://checklist.aou.org/taxa (accessed March 24, 2020).
 The naming of birds in North America has been standardized under the aegis of professional ornithologists in the United States since the late 1800s. It used to be that this list was updated every three or four decades for the first century or so, then it dropped to every decade; now it is updated practically daily, making it a pedant's delight. If you want to be sure what common and scientific name a bird goes by in this country at this moment, you have to go online and check this list. Otherwise, there is a fair chance that you will be wrong.
American Ornithologists' Union. *Check-list of North American Birds.*
 Lancaster, Pennsylvania: American Ornithologists' Union, 1931.
 This edition describes the new *migratory* subspecies of the Bachman's Sparrow.

Audubon, J. J. *Ornithological Biographies.*
Vol. 2. Edinburgh: Adam and Charles Black, 1834.
Here Audubon provides notes on southeastern distribution of Bachman's Sparrow during the early 1830s.

Baker, R. R. *The Evolutionary Ecology of Animal Migration.* New York: Holmes & Meier Publishers, 1978.

Barker, F. K., K. J. Burns, J. Klicka, S. M. Lanyon, and I. J. Lovette. "New insights into New World biogeography: An integrated view from the phylogeny of blackbirds, cardinals, sparrows, tanagers, warblers, and allies." *Auk* 132(2015):333–348.
This paper presents the phylogenetic analysis on which Winger et al.'s Migratory Syndrome hypothesis is based.

Beebe, W. "Avian migration at Rancho Grande in north-central Venezuela." *Zoologica* 32(1947):153–168.
Beebe documents capture of members of several species of tropical rain forest residents at a highland pass in evident intratropical migration.

Bent, A. C. "Life histories of North American birds." *United States National Museum Bulletin* 107, 113, 121, 126, 130, 135, 142, 146, 162, 167, 170, 174, 176, 179, 191, 195–197, 203, 211(1919–1958).
Bent spent the majority of his professional career compiling life history data for all of the breeding species in North America. The result was an extraordinarily comprehensive summary based on everything the author could find on each bird based not only on published information but also on extensive correspondence with the known experts for each species. From our perspective, a key aspect was the near absence of information on birds outside the breeding season and no mention of a postbreeding period.

Berthold, P. "The control of migration in European warblers." *Proceedings of the International Ornithological Congress* 19(1988):215–249.
A fine summary of the experimental basis for the route selection or direction and time hypothesis for how young birds are able to locate their ancestral wintering area.

Berthold, P., E. Gwinner, and E. Sonnenschein, eds. *Avian Migration.* Berlin: Springer, 2003.

Blacklock, G. W. *Checklist of Birds of the Welder Wildlife Refuge.* Sinton, Texas: Welder Wildlife Foundation, 1984.
The document providing baseline information on bird distribution at Welder as of 1984; used for comparison with my census work 2007–2010.

Brosset, A. "Ecological localization of migratory birds in the equatorial forest of Gabon [in French]." *Alauda* 52(1968):81–101.

Brown, J. L. "The evolution of diversity in avian territorial systems." *Wilson Bulletin* 6(1964):160–169.
In this seminal paper, Brown places the phenomenon of territoriality on a sound theoretical footing.

Buechner, H. K., and J. H. Buechner, eds. "The avifauna of northern Latin America." *Smithsonian Contributions to Zoology* 26(1970).
This volume was stimulated by the renowned scientist, diplomat, and conservationist

William Vogt. He worried that rapid destruction of tropical habitats could result in the disappearance of migratory birds, and he hoped that the cadre of leading ornithologists and ecologists assembled for the symposium would help to bring this problem to the attention of the government and the public at large, in vain as it turned out.

Burleigh, T. D. *Georgia Birds*. Norman, Oklahoma: University of Oklahoma Press, 1958.

Carson, Rachel. *Silent Spring*. New York: Houghton Mifflin, 1962.

Chesser, R. T. "Seasonal distribution and ecology of South American austral migrant flycatchers." In *Birds of Two Worlds*, edited by R. Greenberg and P. Marra, 168–181. Baltimore, Maryland: Johns Hopkins University Press, 2005.

Chesser, R. T., and D. J. Levey. "Austral migrants and the evolution of migration in New World birds: Diet, habitat, and migration revisited." *American Naturalist* 152(1998):311–319.

Cooke, W. W. *Bird Migration*. Washington, DC: US Department of Agriculture, 1915. Dr. Cooke's compilation represents a remarkably early attempt to systematize knowledge about hemispheric bird migration.

Cornell Laboratory of Ornithology. *Birds of the World*. https://birdsoftheworld.org/. This work represents a combination of Poole's *The Birds of North America* and del Hoyo et al.'s *Handbook of the Birds of the World*.

Cox, G. W. "The evolution of avian migration systems between temperate and tropical regions of the New World." *American Naturalist*, 126(1985):451–474.

Crawford, R. L. "Bird casualties at a Leon County, Florida TV tower: A 25–year migration study." *Bulletin of the Tall Timbers Research Station*, 22(1981):1–30. Shortly after TV towers became commonplace in the 1960s, ornithologists recognized the extraordinary research opportunity that they represented. Extending as much as two thousand feet into the air, the guy-wires for the tower killed some birds attracted to the tower aviation warning lights under certain weather conditions (low clouds), thereby providing a sample of transients in the middle of a night's migratory journey. These samples serve as incredible snapshots of migration timing by age, sex, and species for a wide range of migratory birds. The Tall Timbers tower was among the most important of these reports for the quality of data, being located along a major fall migration route for a large number of species breeding in eastern North America.

Darwin, C. *On the Origin of Species*. London: J. Murray, 1859. There is little that need be added to the encomia for this ground-breaking work. From the perspective of our consideration, Darwin's discussion of the ability to mold behaviors through the process of artificial selection is especially important.

———. *The Descent of Man, and Selection in Relation to Sex*. London: John Murray, 1871.

Deakin, J. E., C. G. Guglielmo, and Y. E. Morbey. "Sex differences in migratory restlessness behavior in a Nearctic-Neotropical songbird." *Auk* 136(2019)ukz017. https://doi .org/10.1093/auk/ukz017.

del Hoyo, J., A. Elliott, and D. A. Christie. *Handbook of the Birds of the World*. Barcelona: Lynx Edicions, Barcelona, 1992–2014. This work has been combined with Poole (1992–2015), *The Birds of North America*, under the title *Birds of the World*, published online by the Cornell Laboratory of Ornithology (https://birdsoftheworld.org/).

Dingle, H. *Migration: The Biology of Life on the Move*. New York: Oxford University Press, 1996.

Here Dingle takes the concept of Migratory Syndrome, first used with regard to insect movements, and expands it to include migration in other groups including birds.

Dixon, C. *The Migration of Birds*. London: Horace Cox, 1897.

Drake, V. A., and A. G. Gatehouse, eds. *Insect Migration: Tracking Resources through Space and Time*. Cambridge, UK: Cambridge University Press, 1995.

Papers in this volume present the Migratory Syndrome hypothesis for control of insect migration.

Felsenstein, J. "Evolutionary trees from gene frequencies and quantitative characters: Finding maximum likelihood estimates." *Evolution* 35(1981):1229–1242.

Fiorillo, A. R. "Dinosaurs of Alaska: Implications for the Cretaceous origin of Beringia." In *The Terrane Puzzle: New Perspectives on Paleontology and Stratigraphy from the North American Cordillera*, edited by R. B. Blodgett and G. D. Stanley Jr., 313–326. *Geological Society of America, Special Paper 442*, 2008.

Forbush, E. H., and J. R. May. *A Natural History of American Birds of Eastern and Central North America*. New York: Houghton Mifflin, 1955.

For half a century, this volume was the most popular summary of avian life histories in North America. For our purposes, it is notable for its almost complete lack of information on most birds during the eight or nine months comprising the non-breeding period.

Fretwell, S. D. *Populations in a Seasonal Environment*. Princeton, New Jersey: Princeton University Press, 1972.

Many of the ideas on which the concept of migration as a form of dispersal is built came from this little book, including the obvious fact that it is the breeding season for migrants when competition was most relaxed due to superabundance of food.

Gauthreaux, S. A., Jr. "The ecology and evolution of avian migration systems." In *Avian Biology*, vol. 6, edited by D. S. Farner and J. R. King, 93–168. New York: Academic Press, 1982.

Sid Gauthreaux, a student of George Lowery's from Louisiana State University, is a strong proponent of the effects of male dominance on the shaping of bird migration, a view expressed clearly in his comprehensive review of the topic.

Gill, F., and M. Wright. *Birds of the World: Recommended English Names*. Princeton, New Jersey: Princeton University Press, 2006.

This book is the source used for English and scientific names for birds mentioned that are not covered by the American Ornithological Society's checklist.

Gould, S. J., and R. C. Lewontin. "The spandrels of San Marco and the Panglossian paradigm: A critique of the adaptationist programme." *Proceedings of the Royal Society of London. Series B, Biological Sciences* 205(1979):581–598.

This paper has been influential in disparaging the science of field biology.

Gould, S. J., and E. S. Vrba. "Exaptation—a missing term in the science of form." *Paleobiology* 8(1982):4–15.

A remarkable idea that helps explain what Lamarck found inexplicable—namely how individuals can seemingly be "preadapted" for specific changes in their environment.

Groebbels, F. "On the physiology of migratory birds [in German]." *Verhandlungen der Ornithologisch Gesellschaft in Bayern* 18(1928):44–74.
Groebbels first described the basic physiological states associated with migration, *Zugdisposition* and *Zugstimmung*, in this paper.

Guo, P., Q. Liu, Y. Xu, K. Jiang, M. Hou, L. Ding, R. A. Pyron, and F. T. Burbrink. "Out of Asia: Natricine snakes support the Cenozoic Beringian dispersal hypothesis." *Molecular Phylogenetics and Evolution* 63(2012):825–833.

Gutierrez, N. M., L. F. Hinjosa, J. P. LeRoux, and V. Pedroza. "Evidence for an early-middle Miocene age of the Navidad Formation (central Chile): Paleontological, paleoclimatic and tectonic implications." *Andean Geology* 40(2013):66–78.

Gwinner, E. "Circadian and circannual rhythms in birds." In *Avian Biology*, vol. 5, edited by D. S. Farner and J. R. King, 221–285. New York: Academic Press, 1975.
Herein, Dr. Gwinner summarizes his work on the extraordinary discovery of the inherited nature of the timing of migration.

Haug, G. D., and R. Tiedemann. "Effect of the formation of the Isthmus of Panama on Atlantic Ocean thermohaline circulation." *Nature* 393(1998):673–676.

Helbig, A. J. "Evolution of bird migration: A phylogenetic and biogeographic perspective." In *Avian Migration* edited by P. Berthold, E. Gwinner, and E. Sonnenschein, 3–21. Heidelberg: Springer, 2003.

Helm, B., and E. Gwinner. "Migratory restlessness in an equatorial nonmigratory bird." *PLoS Biology* 4(2006):611–614.
Barbara Helm continues the remarkable contributions of the Max Planck Institute with her former advisor and colleague, Dr. Gwinner, with this paper reporting *Zugunruhe* (frustrated migration) in a nonmigratory bird native to the tropics of central Africa.

Hilty, S. L. "Family Thraupidae (tanagers)." In *Handbook of the Birds of the World*, vol. 16, edited by J. del Hoyo, A. Elliott, and D. A. Christie, 46–329. Barcelona: Lynx Edicions, 2011.

Jones, P. "Community dynamics of arboreal insectivorous birds in African savannas in relation to seasonal rainfall patterns and habitat change." In *Dynamics of Tropical Communities*, edited by D. M. Newberry, H. H. T. Prins, and N. D. Brown, 421–427. Oxford: Blackwell, 1999.
Dr. Jones here summarizes the complex migrations tracking seasonal change that occur in Africa.

Karr, J. R. "On the relative abundance of migrants from the north temperate zone in tropical habitats." *Wilson Bulletin* 88(1976):433–458.
This paper provides perhaps the most complete explanation of migrant winter ecology from a Northern Home Theory perspective of migrant ecology and evolution.

Kirby, M. X., D. S. Jones, and B. J. McFadden. "Lower Miocene stratigraphy along the Panama Canal and its bearing on the Central American peninsula." *PLoS ONE* 3(2008):e2791. https://doi.org/10.1371/journal.pone.0002791.

Kricher, J. *Peterson Guide to Bird Behavior*. New York: Houghton Mifflin Harcourt, 2020.

Lack, David L. *The Life of the Robin*. London: Witherby, 1943.

A superb example of the field biologist's art, detailing life history of this European partial migrant.

Levey, D. J., and F. G. Stiles. "Evolutionary precursors to long-distance migration: Resource availability and movement patterns in Neotropical landbirds." *American Naturalist* 140(1992):447–476.

Lockwood, M. W., and B. Freeman. *The Texas Ornithological Society Handbook of Texas Birds*. College Station: Texas A&M University Press, 2004.

This book represents an attempt to bring distribution of Texas birds up to date—based almost entirely on sight records. Distribution summarized in *The Bird Life of Texas* was based on specimens for the most part.

Loss, S. R., and P. P. Marra. "Population impacts of free-ranging domestic cats on mainland vertebrates." *Frontiers in Ecology and the Environment*, 15(2017):502–509. https://doi.org/10.1002/fee.1633.

One of many salvos by anti-cat enthusiasts.

Loss, S. R., T. Will, and P. P. Marra. "The impact of free-ranging domestic cats on wildlife of the United States." *Nature Communications* 4(2013):1396. https://doi.org/10.1038/ncomms2380.

Here Loss and Marra present data on estimated numbers of terrestrial vertebrate deaths due to cats based on extrapolation.

Lowery, G. H. "Trans-Gulf spring migration of birds and the coastal hiatus." *Wilson Bulletin* 57(1945):92–121.

One of several contributions in the long-running battle over the nature of spring versus fall bird migration over (or around) the Gulf of Mexico.

———. "Evidence of trans-Gulf migration." *Auk* 63(1946):175–211.

One of several contributions in the long-running battle over the nature of spring versus fall bird migration over (or around) the Gulf of Mexico.

Lynn, W. S., F. Santiago-Alva, J. Lindenmayer, J. Hadidan, A. Wallach, and B. J. King. "A moral panic over cats." *Conservation Biology* 33(2019):769–776.

An attempt to inject science and civility into the cat conflict.

MacArthur, R. H. *Geographical Ecology*. New York: Harper and Row, 1972.

This slim volume, published shortly before the author's death, provides brief summaries of MacArthur's main theoretical contributions, including those pertaining to migration.

MacArthur, R. H., and E. O. Wilson. *The Theory of Island Biogeography*. Princeton, New Jersey: Princeton University Press, 1967.

The theory proposed by MacArthur and Wilson holds that the number of species on an "island" (an isolated piece of habitat) is determined by the size of the "island" and its distance from potential sources of colonists. The idea has been used to explain how the breaking up of huge areas of forest in the eastern United States into small woodlots could explain population decline in forest-related migrants.

Marra, P. P., and R. L. Holberton. "Corticosterone levels as indicators of habitat quality: Effects of habitat segregation in a migratory bird during the non-breeding season." *Oecologia* 116(1998):284–292.

This paper presents the idea that male superiority in competition for winter food with females results in superior physiological condition that carries over during spring migration and into the breeding season.

Mayr, E. *Animal Species and Evolution.* Cambridge, Massachusetts: Harvard University Press, 1963.

Perhaps the most important statement and explanation of the "biological species" concept in which speciation is understood to require geographic isolation of a population from related populations.

Mayr, E., and W. Meise. "Theories on the history of migration [in German]." *Vogelzug* 1(1930):149–172.

McClure, E. *Migration and Survival of the Birds of Asia.* Bangkok, Thailand: US Army Component, SEATO Medical Research Laboratory, 1974.

Surely one of the more laudatory results of the Vietnam experiment, Dr. McClure's efforts provide outstanding documentary evidence (banding and recapture) of a vast system of intratropical migration throughout Southeast Asia.

Mengel, R. M. "The probable history of species formation in some northern wood warblers." *Living Bird* 3(1964):9–43.

This fascinating paper linked current distribution for related warblers to Pleistocene glaciations. The conclusions are now discredited by improved understanding of both genetics and paleohistory.

Mewaldt, L. R., S. S. Kibby, and M. L. Morton. "Comparative biology of Pacific coastal White-crowned Sparrows." *Condor* 70(1968):14–30.

Mineau, P., F. McKinney, and S. R. Derrickson. "Forced copulation in waterfowl." *Behaviour* 86(1983), Issues 3 and 4.

This review presents the phenomenon of forced copulation entirely from the perspective of male fitness benefits.

Monroe, B. L., Jr. "A distributional survey of the birds of Honduras." *Ornithological Monographs* 7(1968).

This monograph contains information on wintering localities for the endangered Golden-cheeked Warbler.

Moreau, R. E. "The place of Africa in the Palaearctic migration system." *Journal of Animal Ecology* 21(1952):250–271.

Herein Moreau presents his paradox. How can 183 temperate zone species invade tropical African communities each year?

———. *The Palaearctic-African Bird Migration Systems.* New York: Academic Press, 1972.

Morel, G., and F. Bourlière. "Ecological relations of the sedentary and migratory avifauna in a Sahel savannah of lower Senegal [in French]." *Terre et Vie* 4(1962):371–393.

This paper was highly influential in formulation of the "migrant as interloper into the tropics" hypothesis.

Newton, I. *The Migration Ecology of Birds.* London: Academic Press, 2008.

Newton's work represents the most complete compendium of migration knowledge of which I am aware.

Nice, M. M. "On the natural history of the Song Sparrow [in German]." *Journal für Ornithologie* 81(1933):552–595; 82(1934):1–96.

An extraordinary account of the differences in migratory movement for individuals of different age and sex in the Song Sparrow.

———. "Studies in the life history of the Song Sparrow. Volume 1." *Transactions of the Linnean Society of New York* 37(1937).
Remarkable insight into the life history of this partial migrant.

———. "The role of territory in bird life." *American Midland Naturalist* 26(1941):441–487.
Here the author describes the characteristics for the various types of breeding territories seen in birds. The "Type A" territory, that which is seen in most monogamous songbirds, is defined as an exclusive area within which each pair may mate, nest, and feed themselves and their young without interference by others of the same species.

Nolan, V., Jr. "The ecology and behavior of the Prairie Warbler *Dendroica discolor*." *Ornithological Monographs* 26(1978).
Nolan's work presents the most complete knowledge of life history for a long-distance migrant of which I am aware.

Norwine, J., and K. John, eds. *South Texas Climate 2100: The Changing Climate of South Texas 1900–2100*. Kingsville, Texas: Texas A&M University, Kingsville, 2007.
Articles in this book describe climate change underway in the region and its potential effects on various aspects of the environment, including animals and plants.

Oberholser, H. C. *The Bird Life of Texas*. 2 vols. Austin, Texas: University of Texas Press, 1974.
Oberholser spent over forty years compiling information on Texas bird distribution and taxonomy up until his death in 1963. Senior editor Edgar Kincaid, assisted by John Rowlett and Suzanne Winkler, brought these materials up to date, particularly the distributional data, for these two volumes. The result is an extraordinary trove of information, especially with regard to historical changes in distribution.

Peek, F. W. "An experimental study of the territorial function of vocal and visual display in the male red-winged blackbird (*Agelaius phoeniceus*)." *Animal Behaviour* 20(1972):112–118.
An investigation testing the importance of brightly colored plumage in male territorial interactions.

Piersma, T., J. Pérez-Tris, H. Mouritsen, U. Bauchinger, and F. Bairlein. "Is there a migratory syndrome common to all migrant birds?" *Annals of the New York Academy of Sciences* 1046(2005):282–293.
Here Piersma and his coauthors present the extraordinarily useful concept of adaptation versus exaptation for considering migratory bird attributes.

Pigg, K. B., S. M. Ickert-Bond, and J. Wen. "Anatomically preserved *Liquidambar* (Altingiaceae) from the middle Miocene of Yakima Canyon, Washington state, USA, and its biogeographic implications." *American Journal of Botany* 91(2004):499–509.

Pliny, the Elder. *The Natural History*, 77–79 AD. Translation by John Bostock and H. T. Riley. London: George Bell and Sons, 1855.

Poole, A., ed. *The Birds of North America*. Ithaca, NY: American Ornithologists' Union and Cornell Laboratory of Ornithology, 1992–2015.
This work has been combined with del Hoyo et al.'s *Handbook of the Birds of the World* and published online by the Cornell Laboratory of Ornithology under the title *Birds of the World* (https://birdsoftheworld.org/).

Powell, G. V. N., and R. D. Bjork. "Implications of altitudinal migration for conservation strategies to protect tropical biodiversity: A case study of the Resplendent Quetzal *Pharomacrus mocinno* at Monteverde, Costa Rica." *Bird Conservation International* 4(1994):161–174.
This paper represents a remarkable piece of field biology documenting intratropical migration in the quetzal.

Rapp, D. *The Climate Debate.* Self-published, 2012.

Rappole, J. H. "Migrants and space: The wintering ground as a limiting factor for migrant populations." *Bulletin of the Texas Ornithological Society* 7(1974):2–4.
A paper based on my first year of work in Veracruz suggesting that migrant populations could be threatened by loss of wintering-ground habitat.

———. "Intra- and intersexual competition in migratory passerine birds during the nonbreeding season." *Proceedings of the International Ornithological Congress* 17(1988):2308–2317.
A discussion based on extensive field work and examination of museum specimens considering the effects of competition during different parts of the migrant life cycle.

———. *The Ecology of Migrant Birds: A Neotropical Perspective.* Washington, DC: Smithsonian Institution Press, 1995.
This little book represents my first attempt to understand the complexity of the migrant life cycle. Twenty-five years of work by me and many others has resulted in substantial modifications of those ideas.

———. "Not all deaths are equal." *Audubon Naturalist Society Newsletter,* June–July (2005).
A tentative first stab at explaining carrying capacity concepts for migrants during different portions of the annual cycle.

———. *The Avian Migrant.* New York: Columbia University Press, 2013.
An exhaustive treatment of the topic wherein I present my ideas on migration and critically examine alternative hypotheses using a thorough review of the literature.

Rappole, J. H., and K. Ballard. "Passerine postbreeding movements in a Georgia old field community." *Wilson Bulletin* 99(1987):475–480.
Data from this study stimulated the idea of a separate season for migrants after the completion of breeding and before departure on migration, the postbreeding period.

Rappole, J. H., G. W. Blacklock, and J. Norwine. "Apparent rapid range change in South Texas birds: Response to climate change?" In *South Texas Climate 2100: The Changing Climate of South Texas 1900–2100,* edited by J. Norwine and K. John, 131–142. Kingsville: Texas A&M University, Kingsville, 2007.
Using the *Bird Life of Texas* for historical distribution and the *Texas Ornithological Society Handbook of Texas Birds* for current distribution, we documented apparent range change for eighty species of birds in the state.

Rappole, J. H., B. W. Compton, P. Leimgruber, J. Robertson, D. I. King, and S. C. Renner. "Modeling movement of West Nile virus in the Western Hemisphere." *Vector-Borne Zoonotic Diseases* 6(2006):128–139.
This paper models movement of West Nile from its arrival in the New World in 1999 up until 2006, concluding that migratory birds played little role.

Rappole, J. H., S. D. Derrickson, and Z. Hubálek. "Migratory birds and spread of West Nile virus in the Western Hemisphere." *Emerging Infectious Diseases* 6(2000):319–328.

Here we presented predictions for West Nile movement based on observations and hypotheses from the Old World literature.

Rappole, J. H., S. Glasscock, K. Goldberg, D. Song, and S. Faridani. "Range change among New World tropical and subtropical birds." *Bonner Zoologische Monographen* 57(2011):151–167.

This paper documents a remarkable demonstration of the power of dispersal wherein over eighty species of birds extended their breeding ranges north and east by as much as one hundred miles in less than thirty years, presumably in response to creation of favorable habitats by climate change.

Rappole, J. H., B. Helm, and M. A. Ramos. "An integrative framework for understanding the origin and evolution of avian migration." *Journal of Avian Biology* 34(2003):124–128.

Here we respond to the accusation leveled by Dr. Robert Zink that field biology has little to contribute to understanding of the evolution of migration.

Rappole, J. H., and Z. Hubálek. "Migratory birds and West Nile virus." *Journal of Applied Microbiology* 94(2003):47S-58S.

It was in this paper that we first put forward the idea that migratory birds were not the principal culprits in movement of West Nile virus across the Western Hemisphere.

———. "Birds and influenza H5N1 virus movement to and within North America." *Emerging Infectious Diseases* 12(2006):1486–1492.

In this paper we presented data demonstrating the unlikely involvement of migrants in early movement of H5N1, and the low probability that it would ever be found in the New World as a health threat.

Rappole, J. H., and P. Jones. "Evolution of Old and New World migration systems." *Ardea* 90(2002):525–537.

I gave a paper on this topic at a European Ornithological Congress, after which Dr. Jones advised me that marked seasonal changes in rainfall in Africa occurring during north temperate zone "wintering period" made migrant movements a bit more complex than I had presented. We therefore collaborated on this paper, vastly improving on the ideas put forward in my original presentation.

Rappole, J. H., A. H. Kane, R. H. Flores, and A. R. Tipton. "Seasonal variation in habitat use by Great-tailed Grackles in the lower Rio Grande Valley of Texas." *Proceedings of the Animal Damage Control Symposium.* Fort Collins, CO: US Department of Agriculture, 1989.

Here we document the sexual differences in flock composition for grackles during the nonbreeding period.

Rappole, J. H., D. I. King, and J. Diez. "Winter versus breeding habitat limitation for an endangered avian migrant." *Ecological Applications* 13(2003):735–742.

Use of extensive field data on Golden-cheeked Warbler breeding and wintering habitat use, and satellite remote sensing data to quantify actual amounts of each. Results demonstrate that breeding habitat in Texas could accommodate nearly eight times as many birds as wintering habitat can support.

Rappole, J. H., D. I. King, and J. H. Vega Rivera. "Coffee and conservation." *Conservation Biology* 17(2003):334–336.

Here we point out the possible negative consequences of a "shade coffee" campaign for migrants that winter in highland pine-oak habitat of Central America.

Rappole, J. H., and M. V. McDonald. "Cause and effect in population declines of migratory birds." *Auk* 111(1994):652–660.

We use expected observations resulting from assumed breeding versus wintering population limitation for migrants to conclude that most long-distance migrant populations likely are limited by wintering-ground factors, mainly habitat disappearance.

Rappole, J. H., E. S. Morton, T. E. Lovejoy, and J. S. Ruos. *Nearctic Avian Migrants in the Neotropics*. Washington, DC: US Fish and Wildlife Service, 1983.

An early effort at complete summary of literature on New World migrants, along with an assessment of how this knowledge affects our understanding of the group.

Rappole, J. H., A. Pine, D. Swanson, and G. Waggerman. "Conservation and management for migratory birds: Insights from population data and theory for the White-winged Dove." In *Wildlife Science: Linking Ecological Theory and Management Applications*, edited by T. Fulbright and D. Hewitt, 4–20. Gainesville, Texas: CRC Press, 2007.

Here we apply Alan Pine's models for multiple carrying capacity to a long-distance migrant, the White-winged Dove, concluding that winter habitat availability (Central American dry forest) is likely the main contributor to population control.

Rappole, J. H., and M. A. Ramos. "Factors affecting migratory routes over the Gulf of Mexico." *Bird Conservation International* 4(1994):251–262.

One of our repeated efforts to document in the literature fall versus spring route change in a large number of eastern North American migrants.

Rappole, J. H., M. A. Ramos, and K. Winker. "Wintering Wood Thrush movements and mortality in southern Veracruz." *Auk* 106(1989): 402–410.

In this paper we present data documenting survival differences for Wood Thrushes wintering in rain forest as opposed to those found in fields and edges.

Rappole, J. H., and A. R. Tipton. "The evolution of avian migration in the Neotropics." *Ornitología Neotropical* 3(1992):45–55.

Here Alan and I present a model for how dispersal and seasonal climate change could favor evolution of migration to temperate North America by a tropical resident.

Rappole, J. H., and D. W. Warner. "Relationships between behavior, physiology, and weather in avian transients at a migration stopover point." *Oecologia* 26(1976):193–212.

This paper was based mostly on intensive study of a population of transient Northern Waterthrushes stopping over at a Texas mudflat. Birds gained fat and weight rapidly once able to establish temporary feeding territories.

———. "Ecological aspects of avian migrant behavior in Veracruz, Mexico." In *Migrant Birds in the Neotropics: Ecology, Behavior, Conservation, and Distribution*, edited by A. Keast and E. S. Morton, 353–394. Washington, DC: Smithsonian Institution Press, 1980.

Here we present two winter field seasons in Veracruz rain forest documenting season-long territories for twenty species of migrants.

Rappole, J. H., D. W. Warner, and M. A. Ramos. "Territoriality and population structure in a small passerine community." *American Midland Naturalist* 97(1977):110–119.

This little study reports the presence of large numbers of males passing through the breeding territories of several migrant species in a Minnesota bog, documenting that such territories are not maintained for the "exclusive" use of the territory holder and his family, and that food sequestration is an unlikely purpose for migrant breeding territories.

Rettig, T. *Earth Focus and the Science of Climate Change.* Self-published, 2019.
Dr. Rettig presents data documenting climate change but challenging the amount caused by humans.

Ridgely, R. S., T. F. Allnutt, T. Brooks, D. K. McNicol, D. W. Mehlman, B. E. Young, and J. R. Zook. *Digital Distribution Maps of the Birds of the Western Hemisphere*, version 1.0. Arlington, VA: NatureServe, 2003.

Ree, R. H., B. R. Moore, C. O. Webb, and M. J. Donoghue. "A likelihood framework for inferring the evolution of geographic range on phylogenetic trees." *Evolution* 59(2005):2299–2311.

Robbins, C. S., D. K. Dawson, and B. A. Dowell. "Habitat area requirements of breeding forest birds of the Middle Atlantic states." *Wildlife Monographs* 103(1989).

Safriel, U. N. "The evolution of Palearctic migration—the case for southern ancestry." *Israel Journal of Zoology* 41(1995):417–431.
One of the few journal articles supporting the "dominant paradigm."

Schwartz, P. "The Northern Waterthrush in Venezuela." *Living Bird* 3(1964):169–184.
One of the first papers to document a territorial system for a migrant wintering in tropical forest. Dr. Schwartz also reports waterthrush return to winter territories held the previous year.

Sewall, J. O., R. S. W. Van De Wal, K. Van Der Zwan, C. Van Oosterhout, H. A. Dijkstra, and C. R. Scotese. "Climate model boundary conditions for four Cretaceous time slices." *Climate of the Past*, European Geosciences Union 3(2007):647–657. https://doi.org/10.5194/cp-3-647-2007.

Sillet, T. S., and R. T. Holmes. "Variation in survivorship of a migratory songbird throughout its annual cycle." *Journal of Animal Ecology* 71(2002):296–308.
The authors argue that migration is the principal cause of death for adult songbird migrants.

Thomson, A. L. *Problems of Bird-Migration.* London: Witherby, 1926.
Dr. Thompson presents some remarkable insights on the phenomenon long before there were data available to document their accuracy.

Tramer, E. "Proportions of wintering North American birds in disturbed and undisturbed dry tropical habitats." *Condor* 76(1974):460–464.

van Noordwijk, A. J. "Reaction norms in genetical ecology." *BioScience* 39(1989):453–458.
A brilliant paper documenting the widespread nature of variation in phenotypic expression, such as migratory movement, for which there is currently no genotypic explanation.

Van Tyne, J., and A. J. Berger. *Fundamentals of Ornithology.* New York: J. Wiley & Sons, 1959.

Vega Rivera, J. H., W. J. McShea, J. H. Rappole, and C. A. Haas. "Postbreeding movements and habitat use of adult Wood Thrushes in northern Virginia." *Auk* 116(1999):458–466.
This paper, along with the two papers cited below, contain radio-tracking data from

intensive observation of the breeding and postbreeding movements by age and sex in the Wood Thrush.

Vega Rivera, J. H., J. H. Rappole, W. J. McShea, and C. A. Haas. "Pattern and chronology of prebasic molt for the Wood Thrush and its relation to reproduction and migration departure." *Wilson Bulletin* 110(1998):384–392.

———. "Wood Thrush post-fledging movements and habitat use in northern Virginia." *Condor* 100(1998):69–78.

Verhulst, P.-F. "Recherches mathématiques sur la loi d'accroissement de la population." *Nouveau Mémoires de l'Academie Royale des Science et Belles-Lettres de Bruxelles* 18(1845):1–41.

This paper and the one cited immediately below represent the first modeling of the carrying capacity principle, that is, the idea that populations can be limited by amount of available space in organisms that compete for such space (density dependence).

———. "Deuxième mémoire sur la loi d'accroissement de la population." *Nouveau Mémoires de l'Academie Royale des Science et Belles-Lettres de Bruxelles* 20(1847):1–32.

Verner, J. "On the adaptive significance of territoriality." *American Naturalist* 111(1977):769–775.

Vila, R., C. D. Bell, R. Macniven, B. Goldman-Huertas, R. H. Ree, C. R. Marshall, Z. Bálint, K. Johnson, D. Benyamini, and N. Pierce. "Phylogeny and palaeoecology of *Polyommatus* blue butterflies show Beringia was a climate-regulated gateway to the New World." *Proceedings of the Royal Society B: Biological Sciences* 278(2011):2737–2744.

von Haartman, L. "Population dynamics." In *Avian Biology*, vol. 1, edited by D. S. Farner and J. R. King, 391–459. New York: Academic Press, 1971.

Dr. von Haartman's findings that European Pied Flycatcher breeding populations in Finnish pine forest can be increased simply by adding nesting sites (boxes) raises obvious questions regarding the role of food, which was assumed to be the principal determinant of territory number and size at this time.

Ward, P., and A. Zahavi. "The importance of certain assemblages of birds as information centres for food finding." *Ibis* 115(1973):517–534.

This paper on the movement of Queleas from their roosts to and from feeding areas suggests that these birds likely depend on knowledge of flock members for where food can be located.

Whitmore, R. C. "Migrant Birds in the Neotropics: Ecology, Behavior, Distribution, and Conservation by A. Keast, E. S. Morton. A review." *Wilson Bulletin* 93(1981):432–435.

Whitmore's review highlighted the issue of the unknown level of survivorship for migrants of the same species holding wintering territories versus those that were wanderers.

Williams, G. G. "Do birds cross the Gulf of Mexico in spring?" *Auk* 62(1945):98–111.

This paper, along with the others by Dr. Williams cited below, contains the various salvos in the trans-Gulf versus circum-Gulf migration controversy between Dr. Williams (circum-Gulf) and George Lowery and his students (trans-Gulf).

———. "Lowery on trans-Gulf migration." *Auk* 64(1947):217–238.

One of several contributions in the long-running battle over the nature of spring versus fall bird migration over (or around) the Gulf of Mexico.

——. "The nature and causes of the coastal hiatus." *Wilson Bulletin* 62(1950):175–182.
One of several contributions in the long-running battle over the nature of spring versus fall bird migration over (or around) the Gulf of Mexico.

——. "Letter to the editor." *Wilson Bulletin* 63(1951):52–54.
One of several contributions in the long-running battle over the nature of spring versus fall bird migration over (or around) the Gulf of Mexico.

——. "Evolutionary aspects of bird migration." In *Lida Scott Brown Lectures in Ornithology*, 53–85. Los Angeles: University of California, 1958.
Williams riffs here without much data, but some of his musings may turn out to be insightful with regard to our understanding of migration, including the idea that juvenile migrants have information from their parents regarding how to locate proper wintering grounds.

Willis, E. O. "The role of migrant birds at swarms of army ants." *Living Bird* 5(1966):187–231.
A stimulating presentation of the "ecological counterpart" concept.

Wilson, A., and C. L. Bonaparte. *American Ornithology; or, The Natural History of the Birds of the United States*. Philadelphia: Porter and Coates, 1808–1814 [reprinted 1858].
The first comprehensive treatment of North American avifauna.

Wilson, E. O. *Sociobiology*. Cambridge, Massachusetts: Harvard University Press, 1975.
This book seemed to many of us to represent the crest of the wave for dominance of the behavioral ecology discipline in the field of biology during the late twentieth century. The bloom has gone off the rose a bit with the rise of molecular genetics. However, once we are able to figure out exactly how behaviors pass from one generation to the next, combining the knowledge from both fields, I think our understanding of how the world works will take a giant step forward.

Wiltschko, R. "Das Verhalten verfrachteter Vogel." *Vogelwarte* 36(1992):249–310.
In the two papers cited here, along with many others, the Wiltschkos provided experimental evidence on a number of navigational talents of birds, including homing.

Wiltschko, R., and W. Wiltschko. "Mechanisms of orientation and navigation in birds." In *Avian Migration*, edited by P. Berthold, E. Gwinner, and E. Sonnenschein, 433–456. Heidelberg: Springer, 2003.

Winger, B. M., F. K. Barker, and R. H. Ree. "Temperate origins of long-distance seasonal migration in New World songbirds." *Proceedings of the National Academy of Sciences* 111(2014):12115–12120. https://doi.org/10.1073/pnas.1405000111.
A phylogenetic approach to explain the origin of bird migration resulting from the Migratory Syndrome theory.

——. "Temperate origins of long-distance seasonal migration in New World songbirds—Supporting Information." *Proceedings of the National Academy of Sciences* (2014). https://www.pnas.org/content/suppl/2014/07/31/1405000111.DCSupplemental.

Winker, K., P. Escalante, J. H. Rappole, M. A. Ramos, R. Oehlenschlager, and D. W. Warner. "Periodic migration and lowland forest refugia in a sedentary Neotropical bird, Wetmore's Bush-Tanager." *Conservation Biology* 11(1997):692–697.
Documentation of temporary, weather-related altitudinal movement in tropical resident species.

Winker, K., J. H. Rappole, and M. A. Ramos. "The use of movement data as an assay of habitat quality." *Oecologia* 101(1995):211–216.

Density of individuals is demonstrated to be an unreliable indicator of habitat quality.

Winkler, D. "How do migration and dispersal interact?" In *Birds of Two Worlds*, edited by R. Greenberg and P. Marra, 401–413. Baltimore: Johns Hopkins University Press, 2005.

Wooding, S., and R. Ward. "Phylogeography and Pleistocene evolution in the North American Black Bear." *Molecular Biology and Evolution* 14(1997):1096–1105.

Source for the map of Pleistocene forest distribution in Figure 9.2.

Wynn-Edwards, V. C. *Animal Dispersion in Relation to Social Behavior*. London: Oliver & Boyd, 1962.

Wynne-Edwards served as a convenient punching bag for years as a result of his promotion of the concept of "group selection" (individuals acting for the good of the species) as an explanation for many animal behaviors. Research has demonstrated the fallacy of this concept for the most part. Nearly all investigations find that behaviors benefit the individual or, on occasion, their close relatives. However, there is evidence of group selection in social insects and, of course, in humans.

Zink, R. M. "Towards a framework for understanding the evolution of avian migration." *Journal of Avian Biology* 33(2002):433–436.

A succinct presentation of the Migratory Syndrome for evolution of bird migration, and a critique of observationally based ecological theories.

———. "The evolution of avian migration." *Biological Journal of the Linnean Society* 104(2011):237–250.

Zink, R. M., and A. Gardiner. "Glaciation as a migratory switch." *Science Advances* 3(2017):e1603133.

Zink, R. M., and J. Klicka. "The importance of recent ice ages in speciation. A failed paradigm." *Science* 277(1997):1666–1669.

This paper put a giant torpedo below the waterline of the Mengel hypothesis of Pleistocene speciation based on "temperate forest refugia" during recent glacial maxima, demonstrating that evolution of the species involved predated the Pleistocene, based on assumed rates of random genetic variation. Nevertheless, questions remain regarding the speed of genetic change.

Index

My guiding philosophy in constructing this index has been to help readers locate information on their own special interests relevant to the field of migration. Bird species follow the usual rule, listing the group name first, followed by the descriptor, such as "Warbler: Black-and-white." When a multi-word term is usually referenced in its entirety, I list the term alphabetically by first word in the topic, such as "Migratory Syndrome theory of migrant evolution." To avoid redundancies, I do not include words or terms from the appendices or the annotated references.

Page references to photographs and figures are in *italic* type.

39, 61, 154–157, 196, 219–220, 254–256, 263, 268; homing, 33, 169, 180–186, 267; orientation, 29; physiological states, 19–20, 62–63, 67, 88–93, 159, 169–173, 267; route selection, 28–35, 79–88, 181–186, 253–256, 269; seasonal change and, ix, x, 18, 19, 202; sex effects on, x, 27, 28, 263, 268; spring, 77–94; stopover, 88–93, 159; trans-Gulf, 84–88; wanderers (floaters, floating populations), 6, 12–13, 28, 52–58, 62–63, 66, 90, 96, 105–106, 118–119, 147–148, 157, 160, 220, 269; weather effects, ix–x, 5, 14, 17–20, 28, 35–39, 58, 60, 66, 78, 176–177, 186–188, 215–220, 268–269

Migration (or Migratory) Route hypothesis, 182–185, 254. *See also* migration: route selection

Migratory Bird Treaty Act, 222, 239

migratory drop-off, 14, 215–220

migratory restlessness (frustrated movement). See *Zugunruhe*

migratory switch. *See* genetic switch

Migratory Syndrome theory of migrant evolution, 2–5, 16, 96, 134, 168, 172–177, 185, 214–220, 264–267, 289

Miller, Francesca Rappole Wellman, 135

Miller, Olive Beaupré, 186

Mineau, Pierre, 112

Mock, Karilyn, 88

"Modern Synthesis" (of genetics and behavior), 111

monarch butterfly, 184

monogamy. *See* mating system: monogamy

Monroe, Burt, 249–250

Moore, Frank, 92

Moreau, Reginald, 219, 291

Moreau's paradox (of temperate zone migrant semi-annual occupation of tropical communities), 219, 291

Morel, Gerard, 5, 13

Morton, Eugene Siller (Gene), 8, 271

Myers, Peter (Pete), 254

natural selection. *See* selection, natural

Nearctic migrant, 2, 9–12, 20, 58, 96, 104, 252, 268; population decline in, 11, 144–161, 252

Neri, Mara, xvii

Newton, Ian, 86, 176

Newton, Sir Isaac, 217, 273

Nice, Margaret Morse, xvii, 103, 148, 193, 197

niche, x, 2, 6–7, 11–12, 41, 46, 55, 60–61, 96, 104, 127, 147, 174, 177, 200–201, 219, 268–269

Nolan, Val, Jr., xvii, 27, 68, 98, 102, 120–124, 242

nonbreeding territory, 132

North Atlantic route, 32–33, 84

Northern Home theory of migrant evolution, xiii, xvii, 2, 4–7, 11–16, 27, 39, 41, 45, 60–61, 66, 96, 102, 127, 134, 147, 168, 173, 176–177, 216, 219

Norwine, James (Jim), 136, 138

NPR (National Public Radio), 222, 232

Nuthatch, White-breasted, 36, 41, 64, 200

Oberholser, H. C., 138

obligate migrant. *See* calendar migrant

orientation. *See* migration: orientation

origin of migration. *See* migration: evolution of

Oriole: Audubon's, 140; Baltimore, 14, 70, 200; Orchard, 24

ortstreue. *See* homing; site fidelity

Osprey, 34

Ovenbird, 44–45, 49, 51, 53, 57, 82, 118–119

Owl, Great Horned, 243

Parmesan, Camille, 136

partial migration, 170, 193–194, 287. *See also* age differences: in winter range; sexual differences: in winter range

Parula: Northern, 118–119, 209; Tropical, 209

Peek, F. W., 100

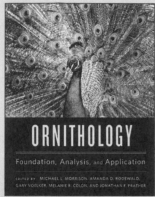
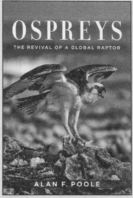